Lecture Notes in Computer Science 11299

Commenced Publication in 1973
Founding and Former Series Editors:
Gerhard Goos, Juris Hartmanis, and Jan van Leeuwen

More information about this series at http://www.springer.com/series/7407

Maria J. Blesa Aguilera · Christian Blum
Haroldo Gambini Santos · Pedro Pinacho-Davidson
Julio Godoy del Campo (Eds.)

Hybrid Metaheuristics

11th International Workshop, HM 2019
Concepción, Chile, January 16–18, 2019
Proceedings

 Springer

Editors
Maria J. Blesa Aguilera ⓘ
Universitat Politècnica de Catalunya
Barcelona, Catalonia

Christian Blum ⓘ
Spanish National Research Council
Bellaterra, Spain

Haroldo Gambini Santos ⓘ
Universidade Federal de Ouro Preto
Ouro Preto, Brazil

Pedro Pinacho-Davidson
Universidad de Concepción
Concepción, Chile

Julio Godoy del Campo
Universidad de Concepción
Concepción, Chile

ISSN 0302-9743   ISSN 1611-3349 (electronic)
Lecture Notes in Computer Science
ISBN 978-3-030-05982-8   ISBN 978-3-030-05983-5 (eBook)
https://doi.org/10.1007/978-3-030-05983-5

Library of Congress Control Number: 2018964128

LNCS Sublibrary: SL1 – Theoretical Computer Science and General Issues

This Springer imprint is published by the registered company Springer Nature Switzerland AG
The registered company address is: Gewerbestrasse 11, 6330 Cham, Switzerland

Preface

After dedicating several decades to the broad development and the deep understanding of sole metaheuristics, it became evident that the concentration on a sole metaheuristic is a rather restrictive strategy when tackling a combinatorial problem. Instead, a skilled combination of concepts from different optimization techniques can provide a more efficient behavior and a higher flexibility for its solvability. These requirements have become especially necessary when dealing with modern real-world and large-scale problems. Hybrid metaheuristics are such techniques for optimization that combine different metaheuristics or integrate AI/OR techniques into metaheuristics.

After more than a decade of worldwide research in this area, today we can be assured that hybrid metaheuristics is a part of experimental science and that its strong interdisciplinarity supports cooperation between researchers with different expertise. Currently, the research on hybrid metaheuristics has reached an interesting point, since it has become clear that many of the optimization problems appearing today require a problem-oriented rather than an algorithm-oriented approach in order to enable a faster and more effective implementation. As a consequence, the hybridization is no longer restricted to different variants of metaheuristics but includes, for example, the combination of mathematical programming, dynamic programming, or constraint programming with metaheuristics, reflecting cross-fertilization in fields such as optimization, algorithmics, mathematical modeling, operations research, statistics, and simulation.

The HM workshops have been consecrated as a forum for researchers from all over the world who study such hybridization strategies and explore the integration of new techniques coming from other areas of expertise. The first edition of HM was held in 2004 and, since then, the event has been held regularly until this 11th edition. Except for its first edition, the proceedings were always published by Springer in this series of *Lecture Notes in Computer Science* (chronologically, volumes LNCS no. 3636, 4030, 4771, 5296, 5818, 6373, 7919, 8457 and 9668).

HM 2019 was the first time that the workshop left Europe and took place in Concepción, Chile, during January 16–18, 2019. This edition was enriched by the presence of four excellent plenary speakers: Haroldo Gambini Santos, from the Universidade Federal de Ouro Preto in Brazil, Manuel López-Ibáñez, from the University of Manchester in the UK, Günther Raidl, from the TU Wien in Austria, and Christian Blum, from the Artificial Intelligence Research Institute (IIIA) in Spain. These researchers are among the leading researchers in the area of hybrid metaheuristics and some of their works make up the state-of-the-art techniques for important optimization problems. We would like to express our gratitude to all of them for having accepted our invitation, and for their participation, which greatly enhanced the quality of the workshop.

On the basis of reviews by the Program Committee members and evaluations by the program chairs, HM 2019 had an acceptance rate of 48% (concerning full papers). We had a double-blind peer review process, with at least three expert referees per manuscript, so that not only originality and overall quality of the papers could be properly evaluated, but also constructive suggestions for improvement could be provided. In light of this, a special thanks is addressed to all the researchers who authored a paper for HM 2019, and to each member of the Program Committee and the external reviewers, who devoted their valuable time and expertise in order to guarantee the scientific quality and interest of this edition of the HM workshops.

January 2019

Maria J. Blesa Aguilera
Christian Blum
Haroldo Gambini Santos
Pedro Pinacho-Davidson
Julio Godoy del Campo

Organization

General Chair

Pedro Pinacho Davidson   University of Concepción, Chile

Program Chairs

Christian Blum   IIIA-CSIC, Spain
Haroldo Gambini Santos   Federal University of Ouro Preto, Brazil

Local Organization Chair

Julio Godoy del Campo   University of Concepción, Chile

Publication and Publicity Chair

Maria J. Blesa   Universitat Politècnica de Catalunya, Catalonia

Posters Evaluation Committee

GIOCo   (Grupo de Investigación en Optimización Combinatorial) Universidad de Concepción, Chile

Program Committee

Felipe Alvelos   Universidade do Minho, Portugal
Claudia Archetti   University of Brescia, Italy
Roberto Asín   Universidad de Concepción, Chile
Francisco Chicano   University of Málaga, Spain
Carlos A. Coello Coello   CINVESTAV-IPN, Mexico
Ricardo Contreras   Universidad de Concepción, Chile
Carlos Cotta   Universidad de Málaga, Spain
Xavier Delorme   ENSM-SE, France
Luca di Gaspero   University of Udine, Italy
Andreas Ernst   Monash University, Australia
Carlos M. Fonseca   University of Coimbra, Portugal
Jin-Kao Hao   University of Angers, France
Laetitia Jourdan   University of Lille 1, Inria Lille, France
Rodrigo Linfanti   Universidad del Bío-Bío, Chile
Manuel López-Ibáñez   University of Manchester, UK
Vittorio Maniezzo   University of Bologna, Italy

Geraldo R. Mateus	Universidade Federal de Minas Gerais, Brazil
Gonzalo Mejía	Universidad de la Sabana, Colombia
Andrés L. Medaglia	Universidad de los Andes, Colombia
Frank Neumann	University of Adelaide, Australia
Gabriela Ochoa	University of Sterling, UK
Luis Paquete	University of Coimbra, Portugal
Jordi Pereira	Universidad Adolfo Ibáñez, Chile
Ma. Angélica Pinninghoff	Universidad de Concepción, Chile
Lorena Pradenas	Universidad de Concepción, Chile
Günther Raidl	Vienna University of Technology, Austria
Helena Ramalhinho-Lourenço	Universitat Pompeu Fabra, Spain
Mauricio G. C. Resende	Amazon.com, USA
Celso C. Ribeiro	Universidade Federal Fluminense, Brazil
María Cristina Riff	Universidad Técnica Federico Santa María, Chile
Alberto Santini	Universitat Pompeu Fabra, Spain
Andrea Schaerf	University of Udine, Italy
Marcone J. Freitas Souza	Universidade Federal de Ouro Preto, Brazil
Marcus Ritt	Universidade Federal do Rio Grande do Sul, Brazil
Thomas Stützle	IRIDIA, Université Libre de Bruxelles, Belgium
Anand Subramanian	Universidade Federal da Paraíba, Brazil
Dhananjay Thiruvady	Monash University, Australia
Thibaut Vidal	Pontifícia Universidade Católica do Rio de Janeiro, Brazil
Stefan Voß	University of Hamburg, Germany

Additional Reviewers

Janniele Araújo
Mario Souto

Contents

Generic CP-Supported CMSA for Binary Integer Linear Programs

Christian Blum[1(✉)] and Haroldo Gambini Santos[2]

[1] Artificial Intelligence Research Institute (IIIA-CSIC), Campus of the UAB,
Bellaterra, Spain
christian.blum@iiia.csic.es
[2] Department of Computer Science, Universidade Federal de Ouro Preto,
Ouro Preto, Brazil
haroldo@ufop.edu.br

Abstract. Construct, Merge, Solve & Adapt (CMSA) is a general hybrid metaheuristic for solving combinatorial optimization problems. At each iteration, CMSA (1) constructs feasible solutions to the tackled problem instance in a probabilistic way and (2) solves a reduced problem instance (if possible) to optimality. The construction of feasible solutions is hereby problem-specific, usually involving a fast greedy heuristic. The goal of this paper is to design a problem-agnostic CMSA variant whose exclusive input is an integer linear program (ILP). In order to reduce the complexity of this task, the current study is restricted to binary ILPs. In addition to a basic problem-agnostic CMSA variant, we also present an extended version that makes use of a constraint propagation engine for constructing solutions. The results show that our technique is able to match the upper bounds of the standalone application of CPLEX in the context of rather easy-to-solve instances, while it generally outperforms the standalone application of CPLEX in the context of hard instances. Moreover, the results indicate that the support of the constraint propagation engine is useful in the context of problems for which finding feasible solutions is rather difficult.

1 Introduction

Construct, Merge, Solve & Adapt (CMSA) [6] is a hybrid metaheuristic that can be applied to any combinatorial optimization problem for which is known a way of generating feasible solutions, and whose subproblems can be solved to optimality by a black-box solver. Moreover, note that CMSA is thought for those problem instances for which the application of the standalone black-box solver is not feasible due to the problem instance size and/or difficulty. The main idea of CMSA is to generate reduced sub-instances of the original problem instances, based on feasible solutions that are constructed at each iteration, and to solve these reduced instances by means of the black-box solver. Obviously, the parameters of CMSA have to be adjusted in order for the size of the

© Springer Nature Switzerland AG 2019
M. J. Blesa Aguilera et al. (Eds.): HM 2019, LNCS 11299, pp. 1–15, 2019.
https://doi.org/10.1007/978-3-030-05983-5_1

reduced sub-instances to be such that the black-box solver can solve them efficiently. CMSA has been applied to several NP-hard combinatorial optimization problems, including minimum common string partition [4,6], the repetition-free longest common subsequence problem [5], and the multi-dimensional knapsack problem [15].

A possible disadvantage of CMSA is the fact that a problem-specific way of probabilistically generating solutions is used in the above-mentioned applications. Therefore, the goal of this paper is to design a CMSA variant that can be easily applied to different combinatorial optimization problems. One way of achieving this goal is the development of a solver for a quite general problem. Combinatorial optimization problems can be conveniently expressed as Integer Linear Programs (ILPs) in the format min $c^T x : Ax = b, x \in \mathbb{Z}^n$, where A indicates a constraints matrix, b and c are the cost and right-hand-side vectors, respectively and x is a vector of decision variables whose values are restricted to integral numbers. In this paper we propose a generic CMSA for binary integer programs (BIPs), that are obtained when $x \in \{0,1\}^n$. This type of problem is generic enough to model a wide range of combinatorial optimization problems, from the classical traveling salesman problem [2] to protein threading problems [19] and a myriad of applications listed in the MIPLIB 2010 collection of problem instances [13]. As CMSA is an algorithm that makes use of a solution construction mechanism at each iteration, one of the challenges that we address in this paper is the fast production of feasible solutions for general BIPs. For this purpose we support the proposed generic CMSA with a constraint propagation (CP) tool for increasing the probability to generate feasible solutions.

This paper is organized as follows. The next section discusses related work. In Sect. 3, the original version of CMSA is presented, which assumes that the type of the tackled problem is known. The generic CMSA proposal for general BIPs is described in Sect. 4. Finally, an extensive experimental evaluation is presented in Sect. 5 and conclusions as well as an outlook to future work are provided in Sect. 6.

2   Related Work

The development of fast, reliable general purpose combinatorial optimization solvers is a topic that occupies operations research practitioners since many years. The main reason being that the structure of real world optimization problems usually does not remain fixed over time: constraints usually change over time and solvers optimized for a very particular problem structure may lose their efficiency in this process. Thus, a remarkable interest in integer linear programming (ILP) software packages exists, with several commercially successful products such as IBM CPLEX, Gurobi and XPRESS. This success can be attributed to the continuous improvements concerning the performance of these solvers [11,12] and the availability of high level languages such AMPL [10]. The application of these solvers to instances with very different structures creates many challenges. From a practical point of view, the most important one is, possibly, the ability of quickly providing high quality feasible solutions: even though a complete search is executed, it is quite often the case that time limits need to be imposed and a truncated search is performed. Thus, several methods have been proposed to try to produce feasible solutions in the first steps of the search process. One of the best known approaches is the so-called *feasibility pump* [8,9].

In the context of metaheuristics, Kochenberger et al. [14] developed a general solver for unconstrained binary quadratic programming (UBQP) problems. A whole range of important combinatorial optimization problems such as set partitioning and k-coloring can be easily modeled as special cases of the UBQP problem. Experiments showed that their general solver was able to produce high quality solutions much faster than the general purpose ILP solver CPLEX for hard problems such as the set partitioning problem. Brito and Santos [18] proposed a local search approach for solving BIPs, obtaining some encouraging results when comparing to the COIN-OR CBC Branch-and-Cut solver. In the context of constraint programming, Benoist et al. [3] proposed a fast heuristic solver (LocalSolver) based on local search. Experiments showed that LocalSolver outperformed several other solvers, especially for what concerns executions with very restricted computation times. In this paper we propose a CMSA solver for solving BIPs. This format is more restricted than the LocalSolver format, where non-linear functions can be used, but much more general than the UBQP, which can be easily modeled as a special case of binary programming. One advantage of BIPs is that several high performance solvers can be used to solve the sub-problems generated within CMSA, a feature that will be explored in the next sections.

3 Original CMSA in the Context of BIPs

As already mentioned, in this work we focus on solving BIPs. Any BIP can be expressed in the following way:

$$\min\{c^T x : Ax \le b, x_j \in \{0,1\} \ \forall j = 1, \ldots, n\} \tag{1}$$

where A is an $m \times n$ matrix, b is the right-hand-size vector of size m, c is a cost vector, and x is the vector of n binary decision variables. Note that m is the number of constraints of this BIP.

In the following we describe the original CMSA algorithm from [6]. However, instead of providing a general description as in [6], our description is already tailored for the application to BIPs. In order to clarify this fact, the algorithm described below is labeled CMSA-BIP. In general, the main idea of CMSA algorithms is to take profit from an efficient complete solver even in the context of problem instances that are too large to be solved directly. The general idea of CMSA is as follows. At each iteration, CMSA probabilistically generates solutions to the tackled problem instance. Next, the solution components that are found in these solutions are added to a sub-instance of the original problem instance. Subsequently, an exact solver is used to solve the sub-instance (if possible in the given time) to optimality.[1] Moreover, the algorithm is equipped with

---

[1] In the context of problems that can be modelled as BIPs, any black-box ILP solver such as, for example, CPLEX can be used for this purpose.

a mechanism for deleting seemingly useless solution components from the sub-instance. This is done such that the sub-instance has a moderate size and can be solved rather quickly to optimality.

In the context of CMSA-BIP, any combination of a variable x_j with one of its values $v \in \{0,1\}$ is a solution component denoted by (x_j, v). Given a BIP instance, the complete set of solution components if denoted by C. Any sub-instance of the given BIP is a subset C' of C, that is, $C' \subseteq C$. Such a sub-instance $C' \subseteq C$ is feasible, if C' contains for each variable x_j $(j = 1, \ldots, n)$ at least one solution component (x_j, v), that is, either $(x_j, 0)$, or $(x_j, 1)$, or both. Moreover, a solution to the given BIP is any binary vector s that fulfills the constraints from Eq. (1). Note that a feasible solution s contains n solution components: $\{(x_j, s_j) \mid j = 1, \ldots, n\}$.

The pseudo-code of the CMSA-BIP algorithm is given in Algorithm 1. At each iteration the following is done. First, the best-so-far solution s^{bsf} is initialized to NULL, indicating that no such solution exists yet. Moreover, sub-instance C' is initialized to the empty set. Note, also, that each solution component $(x_j, v) \in C$ has a so-called age value denoted by $age[(x_j, v)]$. All these age values are initialized to zero at the start of the algorithm. Then, at each iteration, n_a solutions are probabilistically generated in function ProbabilisticSolutionGeneration(C); see line 6 of Algorithm 1. As mentioned above, problem-specific heuristics are generally used for this purpose. The solution components found in the constructed solutions are then added to C'. Next, an ILP solver is applied in function Apply-ILPSolver(C') to find a possibly optimal solution s'_{opt} to the restricted problem instance C' (see below for a more detailed description). Note that NULL is returned in case the ILP solver cannot find any feasible solution. If s'_{opt} is better than the current best-so-far solution s^{bsf}, solution s'_{opt} is taken as the new best-so-far solution. Next, sub-instance C' is adapted on the basis of solution s'_{opt} in conjunction with the age values of the solution components; see function Adapt(C', s'_{opt}, age_{max}) in line 14. This is done as follows. First, the age of all solution components from C' that are not in s'_{opt} is incremented. Moreover, the age of each solution component from s'_{opt} is re-initialized to zero. Subsequently, those solution components from C' with an age value greater than age_{max}—which is a parameter of the algorithm—are removed from C'. This causes that solution components that never appear in solutions derived by the ILP solver do not slow down the solver in subsequent iterations. On the other side, components which appear in the solutions returned by the ILP solver should be maintained in C'.

Finally, the BIP that is solved at each iteration in function ApplyILP-Solver(C') is generated by adding the following constraints to the original BIP. For each $j = 1, \ldots, n$ the following is done. If C' only contains solution component $(x_j, 0)$, the additional constraint $x_j = 0$ is added to the original BIP. Otherwise, if C' only contains solution component $(x_j, 1)$, the additional constraint $x_j = 1$ is added to the original BIP. Nothing is added to the original BIP in case C' contains both solution components. Note that the ILP solver is applied

---

Algorithm 1. CMSA-BIP: CMSA for solving BIPs

1: **given:** a BIP instance, and values for the algorithm parameters
2: $s^{\text{bsf}} := \text{NULL}; C' := \emptyset$
3: $age[(x_j, v)] := 0$ for all $(x_j, v) \in C$
4: **while** CPU time limit not reached **do**
5:     **for** $i = 1, \ldots, n_a$ **do**
6:         $s := \text{ProbabilisticSolutionGeneration}(C)$
7:         **for** $j = 1, \ldots, n$ **do**
8:             **if** $(x_j, s_j) \notin C'$ **then**
9:                 $age[(x_j, s_j)] := 0$
10:                $C' := C' \cup \{(x_j, s_j)\}$
11:            **end if**
12:        **end for**
13:    **end for**
14:    $s'_{\text{opt}} := \text{ApplyILPSolver}(C')$
15:    **if** s'_{opt} is better than s^{bsf} **then** $s^{\text{bsf}} := s'_{\text{opt}}$ **end if**
16:    $\text{Adapt}(C', s'_{\text{opt}}, age_{\text{max}})$
17: **end while**
18: **return** s^{bsf}

---

with a computation time limit of t^{SUB} CPU seconds, which is a parameter of the algorithm.

4   Generic Way of Generating Solutions for BIPs

In those cases in which the optimization problem modeled by the given BIP is not known, we need a generic way of generating solutions to the given BIP in order to be able to apply the CMSA-BIP algorithm described in the previous section. In the following we first describe a basic solution construction mechanims, and afterwards an alternative mechanism which uses a CP tool for increasing the probability to generate feasible solutions. The first algorithm variant is henceforth denoted as GEN-CMSA-BIP (standing for generic CMSA-BIP) and the second algorithm variant as GEN/CP-CMSA-BIP (standing for generic CMSA-BIP with CP support).

4.1   Basic Solution Construction Mechanism

Before starting with the first CMSA-BIP iteration, a node heuristic of the applied ILP solver might be used in order to obtain a first feasible solution. In our case, we used the node heuristic of CPLEX. If, in this way, a feasible solution can be obtained it is stored in s^{bsf}. Otherwise, s^{bsf} is set to NULL. If, after this step, s^{bsf} has value NULL, the LP relaxation of the given BIP is solved. However, in order not to spend too much computation time on this step, a computation time limit of t^{LP} seconds is applied. After this, the possibly optimal solution of the LP relaxation is stored in vector x^{LP}. Then, whenever function

ProbabilisticSolutionGeneration(C) is called, the following is done. First, a so-called sampling vector x^{samp} for sampling new (possibly feasible) solutions by randomized rounding is generated. If $s^{\text{bsf}} \neq$ NULL, x^{samp} is generated based on s^{bsf} and a so-called *determinism rate* $0 < d_{\text{rate}} < 0.5$ as follows:

$$x_j^{\text{samp}} = \begin{cases} d_{\text{rate}} & \text{if } s_j^{\text{bsf}} = 0 \\ 1 - d_{\text{rate}} & \text{if } s_j^{\text{bsf}} = 1 \end{cases}$$

for all $j = 1, \ldots, n$. In case $s^{\text{bsf}} =$ NULL, x_{samp} is—for all $j = 1, \ldots, n$—generated on the basis of x^{LP}:

$$x_j^{\text{samp}} = \begin{cases} x_j^{LP} & \text{if } d_{\text{rate}} \leq x_j^{LP} \leq 1 - d_{\text{rate}} \\ d_{\text{rate}} & \text{if } x_j^{LP} < d_{\text{rate}} \\ 1 - d_{\text{rate}} & \text{if } x_j^{LP} > 1 - d_{\text{rate}} \end{cases}$$

After generating x_{samp}, a possibly infeasible binary solutions s is generated from x_{samp} by randomized rounding. Note that this is done in the order $j = 1, \ldots, n$.

4.2   CP Supported Construction Mechanism

Our algorithm makes use of the Constraint Propagation (CP) engine **cprop** that implements ideas from [1,17].[2] The support of CP is used in the following two ways. First, all constraints are processed and implications derived from the constraint set are detected and the problem is preprocessed to keep those variables fixed throughout the search process. Second, the solution construction mechanism changes in the following way. Instead of deriving values for the variables in the order $j = 1, \ldots, n$, a random order π is chosen for each solution construction. That is, at step j, instead of deriving a value for variable x_j, instead a value for variable $x_{\pi(j)}$ is derived. Then, after deciding for a value for variable $x_{\pi(j)}$, the CP tool checks if this assignment will produce an infeasible solution. If this is the case, variable $x_{\pi(j)}$ is fixed to the alternative value. If, again, the CP tool determines that this setting cannot lead to a feasible solution, the solution construction proceeds as described in Sect. 4.1, that is, finalizing the solution construction without further CP support. Otherwise—that is, if a feasible value can be chosen for the current variable—the CP might indicate possible implications consisting of further variables that, as a consequence, have to be fixed to certain values. All these implications are dealt with, before dealing with the next non-fixed variable according to π.[3]

4.3   An Additional Algorithmic Aspect

Instead of using fixed values for CMSA-BIP parameters d_{rate} and t^{SUB}, we implemented the following scheme. For both parameters we use a lower bound

---

[2] The used CP tool can be obtained at https://github.com/h-g-s/cprop.

[3] Note that, after fixing a value for $x_{\pi(j)}$, the value of $x_{\pi(j+1)}$ might already be fixed due to one of the implications dealt with earlier.

and an upper bound. At the start of CMSA-BIP, the values of d_{rate} and t^{SUB} are set to the lower bound. Whenever an iteration improves s^{bsf}, the values of d_{rate} and t_{SUB} are set back to their respective lower bounds. Otherwise, the values of d_{rate} and t_{SUB} are increased by a factor determined by substracting the lower bound value from the upper bound value and dividing the result by 5.0. Finally, whenever the value of d_{rate}, respectively t_{SUB}, exceeds its upper bound, it is set back to the lower bound value. This procedure is inspired by variable neighborhood search (VNS) [16].

5   Experimental Evaluation

In the following we present an experimental evaluation of GEN-CMSA-BIP and GEN/CP-CMSA-BIP in comparison to the standalone application of the ILP solver IBM ILOG CPLEX v12.7. Note that the same version of CPLEX was applied within both CMSA variants. Moreover, in all cases CPLEX was executed in one-threaded mode. In order to ensure a fair comparison, CPLEX was executed with two different parameter settings in the standalone mode: the default parameter settings, and with the MIP emphasis parameter set to a value of 4 (which means that the focus of CPLEX is on finding good solutions rather than on proving optimality). The default version of CPLEX is henceforth denoted by CPLEX-OPT and the heuristic version of CPLEX by CPLEX-HEUR. All techniques were implemented in ANSI C++ (with the Concert Library of ILOG for implementing everything related to the ILP models), and using GCC 5.4.0 for compiling the software. Moreover, the experimental evaluation was performed on a cluster of PCs with Intel(R) Xeon(R) CPU 5670 CPUs of 12 nuclei of 2933 MHz and at least 40 Gigabytes of RAM.

5.1   Considered Problem Instances

The properties of the 30 selected BIPs are described in Table 1. The first 27 instances are taken from MIPLIB 2010 (http://miplib.zib.de/miplib2010.php), which is one of the best-known libraries for integer linear programming. More specifically, the ILPs on MIPLIB are classified into three hardness categories: *easy*, *hard*, and *open*. From each one of the these categories we chose (more or less randomly) 9 BIPs. In addition, we selected three instances from recent applications found in the literature:

- mcsp-2000-4 is an instance of the minimum common string partition problem with input strings of length 2000 and an alphabet size of four [6]. The hardness of this instance is due to a massive amount of constraints.
- rflcs-2048-3n-div-8 is an instance of the repetition-free longest common subsequence problem with two input strings of length 2048 and an alphabet size of 768 [5]. This instance is hard to solve due to the large number of variables.
- rcjs-20testS6 is an instance of the resource constraint job scheduling problem considered in [7]. Finding feasible solutions for this problem is, for general purpose ILP solvers, rather time consuming.

Table 1. Characteristics of the 30 BIPs that were considered.

BIP instance name	# Cols/Vars	# Rows	Opt. Val.	MIPLIB status
acc-tight5	1339	3052	0.0	Easy
air04	8904	823	56137.0	Easy
cov1075	120	637	20.0	Easy
eilB101	2818	100	1216.92	Easy
ex9	10404	40962	81.0	Easy
netdiversion	129180	119589	242.0	Easy
opm2-z7-s2	2023	31798	−10280.0	Easy
tanglegram1	34759	68342	5182.0	Easy
vpphard	51471	47280	5.0	Easy
ivu52	157591	2116	481.007	Hard
opm2-z12-s14	10800	319508	−64291.0	Hard
p6b	462	5852	−63.0	Hard
protfold	1835	2112	−31.0	Hard
queens-30	900	960	−40.0	Hard
reblock354	3540	19906	−39280521.23	Hard
rmine10	8439	65274	−1913.88	Hard
seymour-disj-10	1209	5108	287.0	Hard
wnq-n100-mw99-14	10000	656900	259.0	Hard
bab1	61152	60680	Unknown	Open
methanosarcina	7930	14604	Unknown	Open
ramos3	2187	2187	Unknown	Open
rmine14	32205	268535	Unknown	Open
rmine25	326599	2953849	Unknown	Open
sts405	405	27270	Unknown	Open
sts729	729	88452	Unknown	Open
t1717	73885	551	Unknown	Open
t1722	36630	338	Unknown	Open
mcsp-2000-4	1335342	4000	Unkonwn	n.a
rflcs-2048-3n-div-8	5461	7480548	Unknown	n.a
rcjs-20testS6	273372	29032	Unknown	n.a

5.2   Parameter Setting

Both generic CMSA variants have the following parameters for which well-working values must be found: (1) the number of solution constructions per iteration (n_a), (2) the maximumm age of solution components (age_{max}), (3) a computation time limit for solving the LP relaxation (t^{LP}), (4) a lower and an upper bound for the determinism rate (d_{rate} (LB) and d_{rate} (UB)), and (5) a lower and an upper bound for the computation time limit of the ILP solver at each iteration (t^{SUB} (LB) and t^{SUB} (UB)).

Concerning age_{max}, it became clear during preliminary experiments that this parameter has not the same importance for GEN-CMSA-BIP and GEN/CP-CMSA-BIP as it has for a problem-specific CMSA. In other words, while a setting of $n_a = 10$ and $age_{max} = 3$ is essentially different to a setting of $n_a = 30$ and $age_{max} = 1$ for a problem-specific CMSA, this is not the case for the generic CMSA variants. This is related to the way of constructing solutions. In a problem-specific CMSA, the greedy heuristic that is used in a probabilistic way biases the search towards a certain area of the search space. This is generally beneficial, but may have the consequence that some solution components that are needed for high-quality solutions have actually a low probability to be included in the constructed solutions. A setting of $age_{max} > 1$ provides age_{max} opportunites—that is, applications of the ILP solver—to find high-quality solutions that incorporate such solution components. In contrast, the way of constructing solutions in the generic CMSA variants does not produce this situtaiton. Therefore, we decided for a setting of $age_{max} = 1$ for all further experiments. Apart from age_{max}, after preliminary experiments we also fixed the following parameter values:

- $n_a = 5$
- $t^{LP} = 10.0$
- The lower bound of t^{SUB} is set to 30.0 and the upper bound to 100.0

The parameter that has the strongest impact on the performance of the generic CMSA variants is d_{rate}. We noticed that both generic CMSA variants are quite sensitive to the setting of the lower and the upper bound for this parameter. However, in order to avoid a fine-tuning for each single problem instance, we decided to identify four representative parameter value configurations in order to cover the characteristics of the 30 selected problem instances. Both generic CMSA variants are then applied with all four parameter configurations to all 30 problem instances. For each problem instance we take the result of the respective best configuration as the final result (and we indicate with which configuration this result was obtained). The four utilized parameter configurations are described in Table 2.

Table 2. The four parameter configurations used for both GEN-CMSA-BIP and GEN/CP-CMSA-BIP.

Parameter configuration	d_{rate} (LB)	d_{rate} (UB)
Configuration 1	0.03	0.08
Configuration 2	0.05	0.15
Configuration 3	0.1	0.3
Configuration 4	0.3	0.5

5.3 Results

All four approaches (GEN-CMSA-BIP, GEN/CP-CMSA-BIP, CPLEX-OPT, and CPLEX-HEUR) were applied with a computation time limit of 1000 CPU seconds to each one of the 30 problem instances. However, as GEN-CMSA-BIP and GEN/CP-CMSA-BIP are stochastic algorithms, they are applied 10 times to each instance, while the two CPLEX variants are applied exactly once to each instance. The numerical results are provided in Table 3, which has the following structure. The first column contains the problem instance name, and the second column provides the value of an optimal solution (if known).[4] The results of GEN-CMSA-BIP and GEN/CP-CMSA-BIP are presented in three columns for each algorithm. The first column (with heading 'Best') contains the best result obtained over 10 runs, the second column (with heading 'Avg.') shows the average of the best results obtained in each of the 10 runs, and the third column indicates the configuration (out of 4) that has produced the corresponding results. Finally, the results of CPLEX-OPT and CPLEX-HEUR are both presented in two columns. The first column shows the value of the best feasible solution produced within the allowed computation time, and the second column shows the best gap (in percent) at the end of each run. Note that a gap of 0.0 indicates that optimality was proven.

The following observations can be made:

- Concerning the BIPs classified as *easy* (see the first nine table rows), it can be noticed that CPLEX-HEUR always generates an optimal solution, even though optimality can not be proven in two cases. GEN/CP-CMSA-BIP–that is, the generic CP-supported CMSA variant—also produces an optimal solution in at least one out of 10 runs for all nine problem instances. However, in three cases, the algorithm fails to produce an optimal solution in all 10 runs per instance. The results of the basic generic CMSA variant (GEN-CMSA-BIP) are quite similar. However, for instance ex9 it is not able to produce any feasible solution, and for instance netdiversion the results are clearly inferior to those of GEN/CP-CMSA-BIP. Nevertheless, the support of CP also comes with a cost. This can be seen when looking at the anytime behaviour of the algorithms as shown for six exemplary cases in Fig. 1. In particular, GEN/CP-CMSA-BIP is often not converging as fast to good solutions as GEN-CMSA-BIP.

- The increased difficulty of the instances labelled as *hard* (see table rows 10-18), produces more differences between the four approaches. In fact, sometimes one of the CPLEX variants is clearly better than the two CMSA versions (see, for example, instance ivu52), and sometimes the generic CMSA variants outperform the CPLEX versions (such as, for example, for instance opm2-z12-s14). As the CP-support is more costly for these instances, the results of GEN-CMSA-BIP are generally a bit better than those of GEN/CP-CMSA-BIP. The effect of the increased cost of the CP support can also nicely

---

[4] Note that all considered BIPs are in standard form, that is, they must be minimized.

Table 3. Numerical results for the 30 considered BIPs.

Instance	Opt. Val.	Gen-Cmsa-Bip			Gen/Cp-Cmsa-Bip			Cplex-Opt		Cplex-Heur	
		Best	Avg.	Conf.	Best	Avg.	Conf.	Value	Gap	Value	Gap
acc-tight5	0.0	0.0	**0.0**	4	0.0	**0.0**	4	**0.0**	0.0	**0.0**	0.0
air04	56137.0	56137.0	**56137.0**	4	56137.0	**56137.0**	4	**56137.0**	0.0	**56137.0**	0.0
cov1075	20.0	20.0	**20.0**	3	20.0	**20.0**	4	**20.0**	0.0	**20.0**	0.0
eilB101	1216.92	1216.92	**1216.92**	4	1216.92	**1216.92**	3	**1216.92**	0.0	**1216.92**	0.0
ex9	81.0	– –	– –	– –	81.0	**81.0**	2	**81.0**	0.0	**81.0**	0.0
netdiversion	242.0	308.0	386.3	3	242.0	288.0	4	276.0	15.5	**242.0**	0.0
opm2-z7-s2	-10280.0	-10280.0	**-10280.0**	4	-10280.0	**-10280.0**	3	**-10280.0**	0.0	**-10280.0**	0.0
tanglegram1	5182.0	5182.0	**5182.0**	4	5182.0	**5182.0**	4	**5182.0**	0.0	**5182.0**	0.0
vpphard	5.0	5.0	**5.0**	3	5.0	5.5	3	**5.0**	0.0	**5.0**	100.0
ivu52	481.007	657.6	884.23	3	16860.0	16860.0	3	647.05	25.77	**490.09**	2.0
opm2-z12-s14	-64291.0	-63848.0	**-62941.5**	3	-64149.0	-61286.2	3	-50200.0	78.62	-44013.0	106.52
p6b	-63.0	-63.0	-62.6	3	-63.0	-61.9	4	-62.0	9.38	**-63.0**	8.08
protfold	-31.0	-29.0	**-27.4**	3	-30.0	-27.0	3	-25.0	43.53	-26.0	54.82
queens-30	-40.0	-39.0	**-38.8**	3	-38.0	-37.7	3	-38.0	83.0	-38.0	83.56
reblock354	-39280521.23	-39261319.4	-39246477.9	4	-39271317.72	-39262125.07	4	**-39277743.28**	0.27	-39259518.82	0.53
rmine10	-1913.88	-1911.39	-1909.05	4	-1911.75	**-1910.72**	3	-1909.29	0.54	-1910.49	0.75
seymour-disj-10	287.0	287.0	**287.5**	3	287.0	287.6	4	288.0	2.08	288.0	2.17
wnq-n100-mw99-14	259.0	259.0	268.2	3	268.0	272.6	4	275.0	13.98	**267.0**	11.92
bab1	?	-218764.89	**-218764.89**	3	-218764.89	**-218764.89**	4	**-218764.89**	1.98	-218764.89	3.13
methanosarcina	?	2730.0	**2735.1**	3	2841.0	2918.2	2	3080.0	56.79	2772.0	56.02
ramos3	?	233.0	236.6	1	232.0	**235.3**	1	248.0	40.99	263.0	44.38
rmine14	?	-4266.31	**-4254.25**	3	-4210.99	-4182.77	3	-962.12	347.84	-1570.21	174.53
rmine25	?	-10297.66	**-8836.25**	3	-1790.65	-1130.70	4	0.0	inf	0.0	inf
sts405	?	342.0	**344.0**	2	343.0	346.2	1	349.0	50.38	348.0	60.06
sts729	?	646.0	**647.90**	1	650.0	651.4	1	661.0	61.25	729.0	65.65
t1717	?	178443.0	**182489.3**	2	188527.0	192958.1	1	200300.0	32.18	192942.0	29.92
t1722	?	120946.0	124910.0	4	119478.0	126627.0	3	119086.0	14.81	**118608.0**	15.24
mcsp-2000-4	?	514.0	**519.5**	2	527.0	527.0	3	527.0	100.0	527.0	100.0
rflcs-2048-3n-div-8	?	-115.0	-105.2	1	-114.0	**-105.3**	4	0.0	inf	0.0	inf
rcjs-20testS6	?	– –	– –	– –	11555.9	**15279.3**	1	20692.4	68.9	– –	– –

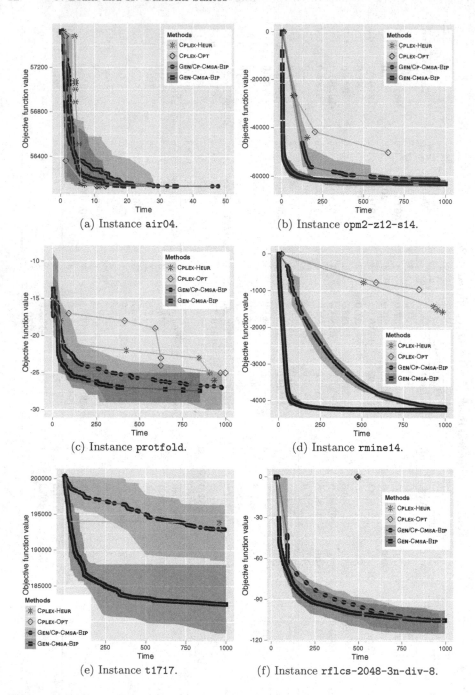

(a) Instance `air04`.

(b) Instance `opm2-z12-s14`.

(c) Instance `protfold`.

(d) Instance `rmine14`.

(e) Instance `t1717`.

(f) Instance `rflcs-2048-3n-div-8`.

Fig. 1. Anytime performance of GEN-CMSA-BIP and GEN/CP-CMSA-BIP in comparison to the two CPLEX variants. The performance of the two generic CMSA versions is shown via the mean performance together with the confidence ribbon (based on 10 independent runs).

be observed in the anytime behaviour of the algorithms for two hard instances in Fig. 1c and b.

- In the context of the nine *open* instances, the generic CMSA variants clearly outperform the standalone application of CPLEX, with the exception of instance t1722. The same holds for the three additional, difficult problem instances (last three table rows). Note, especially, that for instance rcjs-20testS6 the support of CP pays off again, as it is difficult to find feasible solutions for this instance.

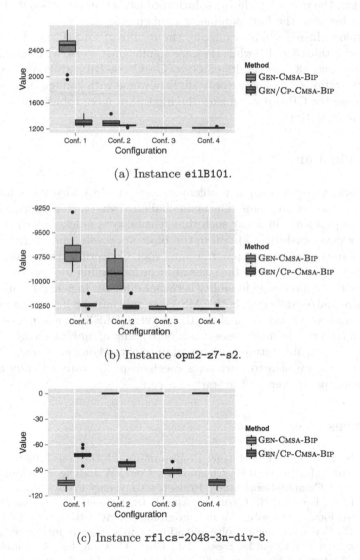

(a) Instance eilB101.

(b) Instance opm2-z7-s2.

(c) Instance rflcs-2048-3n-div-8.

Fig. 2. Boxplots showing the 10 results per algorithm configuration for GEN-CMSA-BIP and GEN/CP-CMSA-BIP in the context of three of the considered problem instances.

Summarizing, with increasing problem size/difficulty, the advantage of the generic CMSA variants over the standalone application of CPLEX becomes more and more pronounced. However, as the cost of the CP support also increases with growing problem size, GEN/CP-CMSA-BIP is only able to outperform GEN-CMSA-BIP when finding feasible solutions is really difficult. However, we noticed that the CP support has also an additional effect, which is shown in the graphics of Fig. 2. Each boxplot shows the final results (obtained by 10 runs per instance) of each of the four parameter configurations for both generic CMSA variants. Interestingly, the use of CP during solution construction flattens out the quality differences between the four parameter configurations. This can be seen in all three boxplots. In Fig. 2b, for example, the results of GEN-CMSA-BIP are good with configurations 3 and 4, while they are significantly worse with configurations 1 and 2. In contrast, the results of GEN/CP-CMSA-BIP, while also being best with configurations 3 and 4, are only slightly worse with configurations 1 and 2. In that sense, the CP-support makes the algorithm more robust with respect to the parameter setting.

6   Conclusions

In this work, we developed a problem-agnostic CMSA algorithm for solving binary linear integer programs. The main challenge was on constructing solutions to unknown problems, in a way such that feasibility is quickly reached. For this purpose we developed—in addition to the basic approach—an algorithm variant that makes use of a constraint programming tool. Concerning the results, we were able to observe that the use of the constraint programming tool pays off for those problems for which reaching feasiblity is rather difficult. In general, with growing problem size and/or difficulty, both CMSA variants have an increasing advantage over the standalone appliaction of the ILP solver CPLEX. In a sense, our generic CMSA can be seen—in many cases—as *a better way of making use of a black-box ILP solver*. Concerning future work, we plan to extend our work towards general ILPs. Moreover, we plan to work on a mechanism for automatically adjusting the algorithm parameters at the start of a run.

References

1. Achterberg, T.: Constraint Integer Programming. Ph.D. thesis (2007)
2. Applegate, D.L., Bixby, R.E., Chvátal, V., Cook, W.J.: The Traveling Salesman Problem: A Computational Study. Princeton University Press, Princeton (2007)
3. Benoist, T., Estellon, B., Gardi, F., Megel, R., Nouioua, K.: LocalSolver 1.x: a black-box local-search solver for 0–1 programming. 4OR **9**(3), 299 (2011)
4. Blum, C.: Construct, merge, solve and adapt: application to unbalanced minimum common string partition. In: Blesa, M.J., Blum, C., Cangelosi, A., Cutello, V., Di Nuovo, A., Pavone, M., Talbi, E.-G. (eds.) HM 2016. LNCS, vol. 9668, pp. 17–31. Springer, Cham (2016). https://doi.org/10.1007/978-3-319-39636-1_2

5. Blum, C., Blesa, M.J.: A comprehensive comparison of metaheuristics for the repetition-free longest common subsequence problem. J. Heuristics **24**, 551–579 (2017)
6. Blum, C., Pinacho, P., López-Ibáñez, M., Lozano, J.A.: Construct, merge, solve & adapt: a new general algorithm for combinatorial optimization. Comput. Oper. Res. **68**, 75–88 (2016)
7. Ernst, A.T., Singh, G.: Lagrangian particle swarm optimization for a resource constrained machine scheduling problem. In: 2012 IEEE Congress on Evolutionary Computation, pp. 1–8 (2012)
8. Fischetti, M., Glover, F., Lodi, A.: The feasibility pump. Math. Program. **104**(1), 91–104 (2005)
9. Fischetti, M., Salvagnin, D.: Feasibility pump 2.0. Math. Program. Comput. **1**(2), 201–222 (2009)
10. Fourer, R., Gay, D., Kernighan, B.: AMPL, vol. 117. Boyd & Fraser Danvers (1993)
11. Gamrath, G., Koch, T., Martin, A., Miltenberger, M., Weninger, D.: Progress in presolving for mixed integer programming. Math. Program. Comput. **7**, 367–398 (2015)
12. Johnson, E., Nemhauser, G., Savelsbergh, W.: Progress in linear programming-based algorithms for integer programming: an exposition. INFORMS J. Comput. **12**, 2–3 (2000)
13. Koch, T., et al.: Miplib 2010. Math. Program. Comput. **3**(2), 103 (2011)
14. Kochenberger, G., et al.: The unconstrained binary quadratic programming problem: a survey. J. Comb. Optim. **28**(1), 58–81 (2014)
15. Lizárraga, E., Blesa, M.J., Blum, C.: Construct, merge, solve and adapt versus large neighborhood search for solving the multi-dimensional knapsack problem: which one works better when? In: Hu, B., López-Ibáñez, M. (eds.) EvoCOP 2017. LNCS, vol. 10197, pp. 60–74. Springer, Cham (2017). https://doi.org/10.1007/978-3-319-55453-2_5
16. Mladenović, N., Hansen, P.: Variable neighborhood search. Comput. Oper. Res. **24**(11), 1097–1100 (1997)
17. Sandholm, T., Shields, R.: Nogood learning for mixed integer programming. Technical report (2006)
18. Souza Brito, S., Gambini Santos, H., Miranda Santos, B.H.: A local search approach for binary programming: feasibility search. In: Blesa, M.J., Blum, C., Voß, S. (eds.) HM 2014. LNCS, vol. 8457, pp. 45–55. Springer, Cham (2014). https://doi.org/10.1007/978-3-319-07644-7_4
19. Xy, J., Li, M., Kim, D., Xu, Y.: RAPTOR: optimal protein threading by linear programming. J. Bioinform. Comput. Biol. **1**(1), 95–117 (2003)

Maximising the Net Present Value of Project Schedules Using CMSA and Parallel ACO

Dhananjay Thiruvady[1(✉)], Christian Blum[2], and Andreas T. Ernst[1]

[1] School of Mathematical Sciences, Monash University, Melbourne, Australia
{dhananjay.thiruvady,andreas.ernst}@monash.edu
[2] Artificial Intelligence Research Institute (IIIA-CSIC),
Campus of the UAB, Bellaterra, Spain
christian.blum@iiia.csic.es

Abstract. This study considers the problem of resource constrained project scheduling to maximise the net present value. A number of tasks must be scheduled within a fixed time horizon. Tasks may have precedences between them and they use a number of common resources when executing. For each resource, there is a limit, and the cumulative resource requirements of all tasks executing at the same time must not exceed the limits. To solve this problem, we develop a hybrid of Construct, Merge, Solve and Adapt (CMSA) and Ant Colony Optimisation (ACO). The methods are implemented in a parallel setting within a multi-core shared memory architecture. The results show that the proposed algorithm outperforms the previous state-of-the-art method, a hybrid of Lagrangian relaxation and ACO.

Keywords: Project scheduling · Net present value
Construct, Merge, Solve & Adapt · Ant Colony Optimisation

1 Introduction

Resource constrained project scheduling is a problem that has been investigated for many years. Due to the complexity of variants of the problem, solving instances with just 100 tasks is still challenging. The details of different project scheduling problems vary, but the basic elements require a number of tasks to be completed with an objective related to the completion time or some value of the tasks. In recent times, maximizing the *net present value* (NPV) of a project has received attention. Each task has a cash flow that may be positive or negative. The aim is to maximize the net present value of the profit, that is, the sum of the discounted cumulative cash flows of the tasks [7, 18, 22–24].

Numerous variants of project scheduling can be found in the literature; see [6] for an early review. All problems considered in [6] consist of tasks, precedences between them, limited shared (renewable) resources and deadlines. The problems are solved with a number of different methods, including heuristics, local

© Springer Nature Switzerland AG 2019
M. J. Blesa Aguilera et al. (Eds.): HM 2019, LNCS 11299, pp. 16–30, 2019.
https://doi.org/10.1007/978-3-030-05983-5_2

search and branch & bound. A discussion on details of methods for solving these problems, including exact approaches, heuristics and meta-heuristics, is provided in [10]. Among meta-heuristics, simulated annealing, genetic algorithms, and tabu search have been applied. Furthermore, heuristic and exact approaches for project scheduling with time windows are discussed in [17].

A closely related problem is resource constrained job scheduling (RCJS) [21], with the main difference being in the objective, i.e., minimising the total weighted tardiness of the tasks. To solve this problem, hybrids of mixed integer programming (MIP) decompositions and meta-heuristics [20], constraint programming and ant colony optimisation (ACO) [21], and parallel implementations of these methods [8,19] have been successfully applied.

For project scheduling with the NPV objective, various heuristic and exact approaches can be found in the literature. Problem instances with up to 98 tasks are solved in [7], with an ACO approach that outperforms other meta-heuristics such as a genetic algorithm, tabu search and simulated annealing. In [18], the authors show that ACO can be effective for a similar problem with up to 50 tasks. A scatter search heuristic described in [23] was shown to be more effective than exact approaches based on branch & bound methods for the same problem [24]. In [13], a hybrid of constraint programming and Lagrangian relaxation is considered which is able to find good feasible solutions easily for this problem. A hybrid of Lagrangian relaxation and ACO [22] was shown to be particularly effective when run in parallel [5].

In this study, we consider resource constrained project scheduling (RCPS) maximising the NPV [14]. This problem is henceforth simply denoted by RCPS-NPV. The problem considers several tasks with varying cash flows, precedences between some of them and a common deadline for all of them. There are also common renewable resources, of which the tasks require some proportion. We solve this problem considering a hybrid of two techniques: (1) Construct, Merge, Solve & Adapt (CMSA) and (2) ACO. Hereby, within the CMSA framework, a parallel implementation of ACO (PACO) is iteratively used to generate solutions for being used within CMSA.

CMSA is a rather recent, generic MIP-based metaheuristic that has shown to be effective on several problems including: the minimum common string partition and minimum covering arborescence problems [4], the repetition-free longest common subsequence problem [3], and unbalanced common string partition [2]. While this method has not yet been tested extensively on a wide range of problems, the results from the initial studies are very promising. At each iteration, CMSA requires a set of promising solutions to be generated by some randomized technique. These solutions are then used to build a reduced sub-instance with respect to the tackled problem instance. This sub-instance is then solved by a complete technique. In the context of problems that can be modelled as MIPs, a general-purpose MIP solver may be used for this purpose. As mentioned above, ACO is used in the case of this paper for generating the set of solutions per iteration. ACO is a meta-heuristic that uses the principles underlying the foraging behavior of ants. This technique has shown to be effective on a number of problems [12]. Project scheduling variants have also been solved with ACO [7,16,18]. The RCPS problem with makespan as the objective was considered, for example, in [16]. Moreover, an ACO for project scheduling where the tasks can execute in multiple modes is described in [7]. In [18], a problem similar to that of [23], but with fewer tasks, was tackled.

The paper is organized as follows. Section 2 provides a MIP model of the RCPS-NPV. Section 3 discusses CMSA and PACO. Section 4 details the experiments conducted and the results obtained from these experiments, before conclusions are provided in Sect. 5.

2   The RCPS-NPV Problem

The RCPS-NPV problem can be formulated as follows. A set $\mathcal{J}$ of n tasks is given, with each task $i \in \mathcal{J}$ having a duration d_i, a cash flow cf_{it} at each time period $t \geq 1$ (which may be positive or negative), and an amount of resource of each type k, r_{ik}, that it requires. While the cash flow of a task may vary over its duration, we can simply calculate the total net value c_i that the task would contribute to the project if it was completed at time 0. From this we can compute the discounted value using discount factor $\alpha > 0$ for start time s_i as $c_i e^{-\alpha(s_i + d_i)}$, where d_i is the duration of task i. Note that the formula $e^{-\alpha t}$ for discounting is equivalent to the commonly used function $1/(1 + \bar{\alpha})^t$ for a suitable choice of α. Let P be the set of precedences P between tasks. We will write $i \to j$ or $(i, j) \in P \subseteq \mathcal{J} \times \mathcal{J}$ to denote that the processing of task i must complete before the processing of j starts.

Given k resources and their limits—that is, $R_1, \ldots, R_k$—the cumulative use of resources by all tasks executing at the same time must satisfy these resource limits. There is also a common deadline for all the tasks, δ, representing the time horizon for completing the project. Without such a deadline, tasks with negative cash flow and no successors would never be completed. Given the objective of maximizing the NPV, the problem can be stated as follows:

$$\max \quad \sum_{i \in \mathcal{J}} c_i e^{-\alpha(s_i + d_i)} \tag{1}$$

$$\text{S.T.} \quad s_i + d_i \leq s_j \qquad\qquad \forall\, (i, j) \in P \tag{2}$$

$$\sum_{i \in S(t)} r_{im} \leq R_m \qquad\qquad \forall\, m = 1, \ldots k \tag{3}$$

$$0 \leq s_i \leq \delta - d_i \qquad\qquad \forall\, i \in \mathcal{J} \tag{4}$$

Hereby, set $S(t)$ consists of tasks executing at time t. The NPV objective function (1) is non-linear and neither convex nor concave, making this problem challenging to solve. Constraints (2) enforce the precedences. Constraints (3) ensure that all the resource limits are satisfied. Constraints (4) require that the deadline is satisfied.

Like in the studies by [14] and [22], the deadline is typically not tight, as the aim is not to minimize the makespan. The available slack means that negative-valued tasks can be scheduled later, thereby slightly increasing the NPV.

2.1   MIP Model

A MIP model for the RCPS-NPV (see also [14]) can be defined as follows: Let $V := \{x_{it} \mid i = 1,\ldots,n \text{ and } t = 1,\ldots,\delta\}$ be a the set of binary variables, where x_{it} takes value 1 if task i completes at time t. The objective is to maximise the NPV:

$$\max \sum_{i \in \mathcal{J}} \sum_{t=1}^{\delta} c_i \, e^{-\alpha t} \, x_{it} \tag{5}$$

$$\text{S.T.} \qquad \sum_{t=1}^{\delta} x_{it} = 1 \qquad \forall \, i \in \mathcal{J} \tag{6}$$

$$\sum_{t=1}^{\delta} t \, x_{jt} - \sum_{t=1}^{\delta} t \, x_{it} \geq d_j \qquad \forall \, (i,j) \in P \tag{7}$$

$$\sum_{i=1}^{n} \sum_{\hat{t}=t}^{t+d_i-1} r_{ik} \, x_{i\hat{t}} \leq R_m \qquad \forall \, m = 1,\ldots,k, \; t \in \{1,\ldots,\delta\} \tag{8}$$

$$x_{it} \in \{0,1\} \qquad \forall \, x_{it} \in V \tag{9}$$

Equation (5) maximises the NPV. Constraints (6) ensure that all tasks are completed exactly once. Constraints (7) ensure that the precedences are satisfied. Constraints (8) require that all the resource limits are satisfied.

For an exact MIP solver this would not be expected to be the most efficient formulation. For the CMSA algorithm described below, an important characteristic of this MIP formulation is that solutions have a low density of non-zeros, just one per activity. Furthermore, instead of solving the full model, in CMSA we only solve small subproblems with a significantly reduced subset of the possible time points. During preliminary experiments it was shown that such type of MIPs can be easily solved by applying a general-purpose MIP solver.

3   The Proposed Algorithm

In this section, we provide the details of our implementation of CMSA. This algorithm learns from sets of solutions which are recombined with the aid of a general-purpose MIP solver. In this study we use ACO, implemented in a parallel setting to ensure that diversity is achieved efficiently as described in [21], to generate these sets of solutions for CMSA at each of its iterations.

3.1   Construct, Merge, Solve and Adapt

Algorithm 1 presents our implementation of CMSA for the tackled RCPS-NPV problem. As input the algorithm takes (1) a problem instance, (2) the number of solutions (n_s), (3) a total computation time limit (t_{total}), (4) a computation time limit per MIP-solver application at each iteration (t_{iter}), and (5) a maximum age limit (a_{max}). CMSA keeps, at all times, a subset V' of the complete set of variables V of the MIP model (see previous section for the definition of V). Moreover, a solution S in the context of CMSA is a subset of V. A solution $S \subseteq V$ is called a valid solution iff assigning value one to all variables in S and zero to the remaining variables in $V \setminus S$, results in a valid RCSP-NPV solution.

The objective function value of a solution S is henceforth denoted by $f(S)$.

Algorithm 1. CMSA for the RCPS-NPV problem

1: INPUT: RCPS-NPV instance, n_s, t_{total}, t_{iter}, a_{max}
2: Initialisation: $V' := \emptyset$, $S^{bs} := \emptyset$, $a_{jt} := 0 \ \forall \ x_{jt} \in V$
3: **while** time limit t_{total} not expired **do**
4:     **for** $i = 1, 2, \ldots, n_s$ **do**   # note that this is done in parallel
5:         $S_i :=$ GenerateSolution()
6:         $V' := V' \cup \{S_i\}$
7:     **end for**
8:     $S^{ib} \leftarrow$ Apply_ILP_Solver(V', S^{bs})
9:     **if** $f(S^{ib}) > f(S^{bs})$ **then** $S^{bs} := S^{ib}$ **end if**
10:    Adapt(V', S^{bs})
11: **end while**
12: OUTPUT: S^{bs}

The algorithm works as follows. First, in line 2, the initial subset of variables (V') is initialized to the empty set. The same is done for the best-so-far solution S^{bs}.[1] Moreover, the age value a_{jt} of each variable $x_{jt} \in V$ is initialized to zero. The main algorithm now executes between Lines 3–11 until the time limit (t_{total}) has expired. At each iteration of CMSA, the following actions are taken. First, a number of n_s solutions is generated in a randomized way, that is, the *Construct* phase of CMSA is executed. In principle, any randomized method to generate feasible solutions is acceptable. However, here we use ACO (see Sect. 3.2). In particular, n_s ACO colonies are run in parallel, where one colony is seeded with the best-so-far solution S^{bs}. The remaining colonies are seeded with a random solution. This method ensures that good parts of the search space and also sufficiently diverse parts of the search space are covered.

Once the solutions have been produced, the variables they contain (remember that these are the variables with value one in the corresponding MIP solutions) are added to V' in line 6 (*Merge* phase of CMSA). Based on V', a reduced MIP model is generated (see below for a detailed description) and then solved by applying a general-purpose MIP solver in function Apply_ILP_Solver(V', S^{bs}) (see line 8 of Algorithm 1). This phase is the *Solve* phase of CMSA. Note that the MIP solver is warm-started with the best-so-far solution S^{bs}. Finally, after potentially updating the best-so-far solution S^{bs} with solution S^{ib} obtained as

---

[1] Note that, for consistency, the objective function value of an empty solution is defined as $-\infty$.

output from the MIP solver, the *Adapt* phase of CMSA is executed in function Adapt(V', S^{bs}) (see line 10) as follows. All variables from V' which are not present in S^{bs} have their age values incremented. The age values of those variables, whose corresponding age values have passed a_{max}, are re-initialized to zero. Moreover, they are removed from V'. Finally, the output of CMSA is solution S^{bs}.

The Restricted MIP Model. The restricted MIP model that is solved in function Apply_ILP_Solver(V',S^{bs}) of Algorithm 1 is obtained from the original MIP model outlined in Sect. 2 as follows. Instead of considering all variables from V, the model is restricted to the variables from V'. In this way the model becomes much smaller and can be relatively easily solved. However, note that the search space of such a restricted model is only a part of the original search space. Only when V' is very large (this may happen when n_s is large and the solution construction is very much random) the restricted MIP may not be solved within the given time. As mentioned above, a time limit (t_{iter}) is used to ensure the solver always terminates sufficiently quickly even if it does not find an optimal solution to the restricted MIP within the given time.

It should be noted that this Merge phase is guaranteed to always produce a solution that is at least as good as any of the solutions that has contributed to V'. This is because (1) all variables that take value one in solution S^{bs} are always present in V', and (2) solution S^{bs} is used for warm-starting the MIP solver. Hence the merge phase provides strong intensification. Diversification relies entirely on the solution generation mechanism. In previous papers (see, for example, [2]), solution generation was based on randomized greedy heuristics. It is expected that using ACO here with a guided—respectively, biased—random solution generation will both reduce the number of variables in the restricted MIPs (as solutions are expected to be of higher quality and, therefore, probably with more parts in common), and will assist the method to better explore the neighborhood of the best-so-far solution. The high-quality results that we will present in Sect. 4 seem to support this hypothesis. However, a deeper comparison between a more randomized solution construction and the construction of solutions by parallel ACO colonies is mandatory for future work.

3.2 Parallel Ant Colony Optimisation

ACO was proposed in [11] to solve combinatorial optimisation problems. The inspiration of these algorithms is the ability of natural ant colonies to find short paths between their nest and food sources.

For the purpose of this work, we use the ACO model for the resource constrained job scheduling problem originally proposed in [21]. This approach was extended to a parallel method in a multi-core shared memory architecture by [5]. For the sake of completeness, the details of the ACO implementation are provided here.

A solution in the ACO model is represented by a permutation of all tasks (π) rather than by the start times of the tasks. This is because there are potentially

---

Algorithm 2. ACO for the RCPS-NPV problem

1: **input**: An RCPS-NPV instance, $\mathcal{T}$, π^{bs} (optional)
2: Initialise π^{bs} (if given as input, otherwise not)
3: **while** termination conditions not satisfied **do**
4:     **for** $j = 1$ to n_{ants} **do:**
5:         $\pi^j := \mathsf{ConstructPermutation}(\mathcal{T})$
6:         $\mathsf{ScheduleTask}(\pi^j)$
7:     **end for**
8:     $\pi^{ib} := \arg\min_{j=1,\ldots,n_{\text{ants}}} f(\pi^j)$
9:     $\pi^{bs} := \mathsf{Update}(\pi^{ib})$
10:    $\mathsf{PheromoneUpdate}(\mathcal{T}, \pi^{bs})$
11:    $cf := \mathsf{ComputeConvergence}(\mathcal{T})$
12:    **if** $cf = \text{true}$ **then** *initialise* $\mathcal{T}$ **end if**
13: **end while**
14: **output**: π^{bs} (converted into a CMSA solution)

---

too many parameters if the ACO model is defined to explicitly learn the start times of the tasks. Given a permutation, a serial scheduling heuristic (see [22]) can be used to generate a resource and precedence feasible schedule consisting of starting times for all tasks in a well-defined way. This is described in Sect. 3.3 below. For the moment it is enough to know that a CMSA solution S (in terms of a set of variables) can be derived from and ACO solution π in a well-defined way. As in the case of CMSA solutions, the objective function value of an ACO solution π is denoted by $f(\pi)$.

The pheromone model of our ACO approach is similar to the one used by [1], that is, the set of pheromone values ($\mathcal{T}$) consist of values τ_{ij} that represent the desirability of selecting task j for position i in the permutations to be built. Ant colony system (ACS) [12] is the specific ACO-variant that was implemented.

The ACO algorithm is shown in Algorithm 2. An instance of the problem and the set of pheromone value $\mathcal{T}$ are provided as input. Additionally, a solution (π^{bs}) can be provided as input which serves the purpose of initially guiding the search towards this solution. If no solution is provided, S^{bs} is initialised to be an empty solution.

The main loop of the algorithm at lines 3–13 runs until a time or iteration limit is exceeded. Within the main loop, a number of solutions (n_{ants}) are constructed ($\mathsf{ConstructPermutation}(\mathcal{T})$). Hereby, a permutation π is built incrementally from left to right by selecting, at each step, a task for the current position $i = 1, \ldots, n$, making use of the pheromone values. Henceforth, $\hat{\mathcal{J}}$ denotes the tasks that can be chosen for position i, that is, $\hat{\mathcal{J}}$ consists of all tasks (1) not assigned already to an earlier position of π and (2) whose predecessors are all already scheduled. In ACS, a task is selected in one of two ways. A random number $q \in (0, 1]$ is generated and a task is selected deterministically if $q < q_0$. That is, task k is chosen for position i of π using:

$$k = \arg\max_{j \in \hat{\mathcal{J}}} \tau_{ij} \tag{10}$$

Otherwise, a probabilistic selection is used where job k is selected according to:

$$P(\pi_i = k) = \frac{\tau_{ik}}{\sum_{j \in \hat{J}} \tau_{ij}} \qquad (11)$$

After each step, a local pheromone update is applied to the selected task k at position i:

$$\tau_{ik} := \tau_{ik} \times \rho + \tau_{min} \qquad (12)$$

where $\tau_{min} := 0.001$ is a limit below which a pheromone value does not drop, so that a task k may always have the possibility of being selected for position i.

After the construction of n_{ants} solutions, the iteration-best solution π^{ib} is determined (line 8) and the global best solution π^{bs} is potentially updated in function $\mathsf{Update}(\pi^{ib})$: $f(\pi^{ib}) > f(\pi^{bs}) \Rightarrow \pi^{bs} := \pi^{ib}$). Then, all pheromone values from $\mathcal{T}$ are updated using the solution components from π^{bs} in function $\mathsf{PheromoneUpdate}(\pi^{bs})$:

$$\tau_{i\pi(i)} = \tau_{i\pi(i)} \cdot \rho + \delta \qquad (13)$$

where $\delta = 0.01$ is set to be a (small) reward. The value of the evaporation rate—that is, $\rho = 0.1$—is the same as the one chosen in the original study [22].

Finally, note that [5] showed different ways of parallelising ACO, where the obvious parallelisation involves building solutions concurrently. For two reasons we chose not to use this method for CMSA, but rather to run multiple colonies. The first, and most significant, is that multiple solutions obtained by a single colony can often consist of very similar components, especially when ACO is close to convergence. Second, the cores available to the algorithm are not fully utilised when the other components of ACO are executing (e.g. the pheromone update). Hence, in our CMSA algorithm we use n_s colonies in parallel, which ensures diversity (especially when initialised with different random solutions) and fully utilises the available cores.

3.3 Scheduling Tasks

As mentioned before, given a feasible permutation π of all tasks, a feasible solution in which the starting times of all tasks are explicitly given can be derived in a well-defined way. We briefly discuss this method in the following. For complete details of this procedure, we refer the reader to [22].

Consider a permutation π. In the given order, sets of tasks are selected where each set consists of a task and its successors. Each set is either net positive-valued or net negative-valued, that is, the total NPV of tasks in the set lead to positive or negative values. Those sets which are positive valued are greedily scheduled at the beginning of the horizon, satisfying the precedences and resources.[2] Tasks in the sets that are negative-valued are scheduled starting at the end of the horizon and working backwards while tasks remain. The motivation for this method is

---

[2] Tasks are selected which have no preceding tasks first. This is followed by selecting dependent tasks that a free to be scheduled. This continues until all tasks have been scheduled.

that positive-valued tasks being scheduled early lead to increasing the NPV, and scheduling negative-valued tasks at the end of the horizon leads to the least decrease in the NPV.

4   Experiments and Results

CMSA was implemented in C++ and compiled with GCC-5.2.0. Gurobi 8.0.0[3] was used to solve the MIPs and ACO's parallel component was implemented using OpenMP [9]. Monash University's Campus Cluster, MonARCH,[4] was used to carry out all experiments. Each machine of the cluster provides 24 cores and 256 GB RAM. Each physical core consists of two hyper-threaded cores with Intel Xeon E5-2680 v3 2.5 GHz, 30 M Cache, 9.60 GT/s QPI, Turbo, HT, 12C/24T (120 W).

We used the same instances as [14]. These problem instances were originally obtained from the PSPLIB [15].[5] These instances are characterized by the number of tasks $\{30, 60, 90, 120\}$, the resource factor $\{0.25, 0.5, 0.75, 1.0\}$, the network complexity $\{1.5, 1.8, 2.1\}$, and the resource strength. Instances with 30, 60 and 90 tasks have resource strengths from the range $\{0.2, 0.5, 0.7, 1.0\}$ while the instances with 120 tasks have resource strengths from $\{0.1, 0.2, 0.3, 0.4, 0.5\}$. The resource factor indicates the proportion of the total amount of each resource that is used, on average, by a task. The resource strengths specify how scarce the resources are with low values indicating tight resource constraints. The network complexity indicates the proportion of precedences, where large values imply a large number of precedences. The benchmark set consists, for each combination of the four parameters indicated above, of 10 instances. This means that there are 480 instances with 30, 60, and 90 tasks, and 600 instances with 120 tasks. In total the benchmark set contains 2040 instances.[6]

The study by [14] was used as a guide to obtain the deadlines and cash flows. For each task, the latest start time (l_i) is determined as $l_j \geq d_i \; \forall i \to j$, and then $\delta := 3.5 \times max_j l_j$. A task's cash flow c_i is generated uniformly at random from $[-500, 1000]$. The discount rate is determined as $\alpha := \sqrt[52]{1 + 0.05} - 1$.

For all runs we allowed the use of 5 cores, for both the MIP solver (Gurobi) and the ACO component (resulting in 5 parallel colonies). The parameters settings for each individual colony were the same as those used in [22]. Additionally, the current state-of-the-art algorithm from [22], which is a hybrid between Lagrangian Relaxation and ACO (henceforth denoted as LRACO), was re-run for all the instances to allow a fair comparison. Moreover, note that ACO also used 5 cores in parallel.

---

[3] http://www.gurobi.com/.

[4] https://confluence.apps.monash.edu/display/monarch/MonARCH+Home.

[5] https://www.sciencedirect.com/science/article/abs/pii/S0377221796001701.

[6] Gurobi can solve most of the problem instances with 30 tasks (the optimal solutions are provided in PSPLIB), a number of instances with 60 tasks (<60%), and a very small proportion of the instances with 90 tasks (<2%). None of the instances with 120 tasks can be solved with Gurobi.

The tuning of the parameters related to CMSA is described in Appendix A. Both algorithms were run once on each problem instance and allowed 15 min of wall-clock time.

4.1   Comparison: CMSA Versus LRACO

Table 1 shows a summary of the comparison between LRACO and CMSA. Averaged over the all the instances with the same number of tasks, the table provides gap ($\frac{UB-LB}{UB}$) for each algorithm and the number of times (out of 480, resp. 600) each one finds the best solution (# best). Moreover, the last table column contains the factor of how many more best solutions were found by CMSA relative to LRACO. For example, in the case of instances with 30 tasks, CMSA found 20.92 times more best solutions than LRACO.

For CMSA, the results are reported as the gap between the CMSA lower bound and the upper bound of LRACO (since we do not have a valid upper bound from CMSA). We see that CMSA performs—on average—better than LRACO for each problem size. Additionally, the number of best solutions found shows that CMSA is significantly more effective. Interestingly, the factor (last table column) drops with an increase in the number of tasks (from 20.92 for 30 tasks compared to 2.92 for 120 tasks). This means that, while CMSA is very strong for rather small problem instances, it still significantly outperforms LRACO on large problem instances. The drops of the performance of CMSA with growing problem size, might be related to the number of iterations that can be performed within the allowed computation time. More specifically, CMSA can generally perform more than 100 iterations within the allowed CPU time for instances with 30 tasks, but generally less than 20 iterations for 120 task problems.

These results are visualized in Fig. 1. The boxplots in this graphic show the improvement of CMSA over LRACO in percent ($\frac{(CMSA_i - LR_i)*100}{LR_i}$, $\forall i \in I$)[7] for subsets of the complete set of instances. A value above 0.00 indicates that CMSA

Table 1. Average gaps of LRACO and CMSA to the upper bound obtained by LRACO computed as $\frac{UB-LB}{UB}$. The number of instances for which an algorithm finds the best solution is shown in column # best, while *Factor* indicates how many times more best solutions were found by CMSA relative to LRACO.

Tasks	LRACO			CMSA			Factor
	Mean	SD	# best	Mean	SD	# best	
30	0.0189	0.0605	12	0.0180	0.0692	251	20.92
60	0.0148	0.0125	25	0.0133	0.0118	369	14.76
90	0.0147	0.0105	59	0.0136	0.0103	355	6.02
120	0.0176	0.0099	153	0.0165	0.0105	446	2.92

---

[7] I is the set of instances.

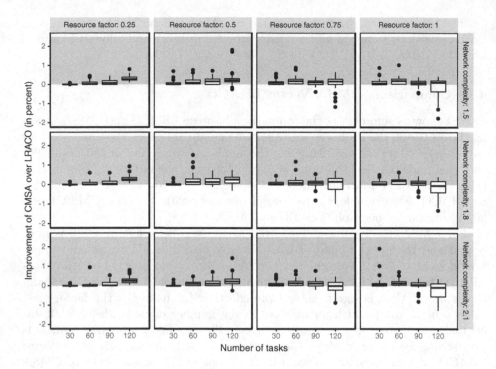

Fig. 1. The improvement of CMSA over LRACO (in percent) for subgroups of instances (as determined by the network complexity and the resource factor). Note that CMSA improves over LRACO whenever a data point has a positive value (that is, when it is located in the area with light blue background). (Color figure online)

performs better than LRACO for the respective instance, while a value below 0.0 indicates the opposite.

In can be observed that for instances with resource factors 0.25 and 0.5, CMSA generally improves over LRACO. However, with growing instance size and resource factor, this advantage of CMSA diminishes. In fact, for instances with 120 tasks and a resource factor of 1.0, LR-ACO seems to slightly outperform CMSA. Note also that a single instance of 30 tasks was removed from this graphic, as the gap was very large in favor of CMSA ($gap = 2.98$). This instance is interesting since it consists of many more negative valued tasks than other instances, leading to solutions with large negative NPVs. In this case CMSA is able to handle the negative tasks much more effectively than LRACO.

Table 2 shows an even more detailed breakdown of the results with respect to the four problem instance parameters (number of tasks, network complexity, resource factor and resource strength). In addition to the observations mentioned earlier, there are a few cases where large differences between the algorithms can be seen. For instances with 30 tasks, with a high resource factor (RF = 1.0) and a relatively high resource strength (RS = 0.7), the difference between CMSA and

Table 2. A breakdown of the results with respect to instance parameters (number of tasks, NC - network complexity, RF - resource factor, RS- resource strength). Gap% shows the same information as the boxplots in Fig. 1 (percentage improvement of CMSA over LRACO).

		30 tasks		60 tasks		90 tasks		120 tasks	
		Gap%	SD	Gap%	SD	Gap%	SD	Gap%	SD
NC	1.5	0.08	0.14	0.16	0.19	0.12	0.17	0.17	0.54
	1.8	0.06	0.11	0.17	0.24	0.11	0.20	0.10	0.39
	2.1	0.07	0.20	0.13	0.16	0.11	0.19	0.07	0.38
RF	0.3	0.01	0.02	0.07	0.13	0.11	0.13	0.30	0.15
	0.5	0.05	0.10	0.18	0.24	0.19	0.18	0.25	0.30
	0.8	0.10	0.13	0.18	0.21	0.11	0.18	0.09	0.57
	1.0	2.61	27.20	0.18	0.18	0.05	0.21	−0.19	0.47
RS	0.1	-	-	-	-	-	-	−0.05	0.80
	0.2	0.16	0.19	0.32	0.26	0.13	0.30	0.24	0.17
	0.3	-	-	-	-	-	-	0.03	0.44
	0.4	-	-	-	-	-	-	0.14	0.25
	0.5	0.08	0.20	0.21	0.16	0.21	0.13	0.20	0.14
	0.7	2.52	27.21	0.08	0.07	0.11	0.08	-	-
	1.0	0.00	0.00	0.00	0.00	0.00	0.00	-	-

LRACO is very high. On the other side, when the instances consist of relatively trivial constraints (high resource strength, i.e., $RS = 1.0$), both algorithms are nearly equal. Interestingly, for a number of very hard instances, LRACO outperforms CMSA ($RF = 1.0$, $RS = 0.1$). As mentioned above, this is likely to be due to the insufficient number of iterations completed the CMSA within the time-limit for such large problems.

5   Conclusion

In this study, we investigate a parallel hybrid of CMSA and ACO for resource constrained project scheduling under the maximization of the net present value. The proposed hybrid algorithm has several key characteristics. First, it makes use of a general purpose MIP solver in an iterative way, in order to solve reduced subinstances of the original problem instances. Second, it makes an efficient use of resources in parallel. This is achieved by running a number of ACO colonies in parallel leading to a diverse solutions space in the considered reduced subinstances. The results show that the hybrid CMSA approach is efficient at solving the problem, and improves upon LRACO, which is currently the state-of-the-art method in the literature.

This study has shown that if a problem can be efficiently modeled both with ACO and with MIP techniques, a generic hybrid can be developed that can solve such a problem efficiently. We are currently working in this direction by developing a generic heuristic that can be applied to a range of problems in a very similar fashion. Furthermore, given that multi-core architectures are readily available nowadays, the parallelisation is straightforward.

A further line of investigation is related to developing a tighter coupling between ACO and CMSA. Since the MIP leads the search towards local optima, the solution information could be used to further assist ACO. For example, the solutions found by the MIPs could be used to generate or update the pheromone matrices at each iteration.

A key component of CMSA is defining an efficient MIP model. In this study, a straightforward MIP model has been used, however, the results could significantly improve with more efficient models. For example, the cumulative model proposed in [22] is more efficient, but requires a different way of extracting solution information. For this purpose, we are developing an alternative method labeled *Merge search*.

Acknowledgements. This research was supported in part by the Monash eResearch Centre and eSolutions-Research Support Services through the use of the MonARCH HPC Cluster.

A   CMSA Parameter Selection

In order to determine the parameter settings of CMSA for the experiments, a subset of the problem instances were selected and used for testing. The parameters of interest are the MIP time limit (we considered either 60 or 120 s), the number of ACO iterations (500, 1000 and 2000 iterations) and the maximum age limit (3, 5 and 10). The instances selected consist of 60 (small) and 120 (large) tasks, with resource factors of 0.5 (low-medium) and 1.0 (high), resource strengths of 0.5 (low-medium) and 1.0 (high). Each run was given 5 ACO colonies (and hence 5 cores) and 15 min of wall-clock time.

Table 3. Number of best solutions found by varying parameters (MIP time limit - MIP TL., ACO iterations - ACO iter. and Age limit - Age).

Param.	Value	60 tasks	120 tasks
MIP TL.	60	**4**	**11**
	120	2	1
ACO Iter.	500	**4**	4
	1000	0	4
	2000	2	4
Age	3	**3**	4
	5	1	**7**
	10	**3**	1

Table 3 shows the number of instances for which the best solution was found with the respective settings. We consider the MIP time limit first, choose the best option, then select the best value for the number of ACO iterations, choose the best option, and finally select between the age limits. For the MIP time limit, 60 s is best for 60 and 120 tasks. Comparing iterations for 60 tasks, 500 is best and we use the same value for 120 tasks observing that there is no advantage. The best age limit for 120 tasks is 5, but because there is no obvious advantage for 60 tasks, we pick 3 in order for the built MIPs to be smaller and the MIP solver, therefore, faster.

References

1. den Besten, M., Stützle, T., Dorigo, M.: Ant colony optimization for the total weighted tardiness problem. In: Schoenauer, M., et al. (eds.) PPSN 2000. LNCS, vol. 1917, pp. 611–620. Springer, Heidelberg (2000). https://doi.org/10.1007/3-540-45356-3_60
2. Blum, C.: Construct, merge, solve and adapt: application to unbalanced minimum common string partition. In: Blesa, M.J., et al. (eds.) HM 2016. LNCS, vol. 9668, pp. 17–31. Springer, Cham (2016). https://doi.org/10.1007/978-3-319-39636-1_2
3. Blum, C., Blesa, M.J.: Construct, merge, solve and adapt: application to the repetition-free longest common subsequence problem. In: Chicano, F., Hu, B., García-Sánchez, P. (eds.) EvoCOP 2016. LNCS, vol. 9595, pp. 46–57. Springer, Cham (2016). https://doi.org/10.1007/978-3-319-30698-8_4
4. Blum, C., Pinacho, P., López-Ibáñez, M., Lozano, J.A.: Construct, merge, solve & adapt a new general algorithm for combinatorial optimization. Comput. Oper. Res. **68**, 75–88 (2016)
5. Brent, O., Thiruvady, D., Gómez-Iglesias, A., Garcia-Flores, R.: A parallel lagrangian-ACO heuristic for project scheduling. In: 2014 IEEE Congress on Evolutionary Computation (CEC), pp. 2985–2991 (2014)
6. Brucker, P., Drexl, A., Möhring, R., Neumann, K., Pesch, E.: Resource-constrained project scheduling: notation, classification, models, and methods. Eur. J. Oper. Res. **112**, 3–41 (1999)
7. Chen, W.N., Zhang, J., Chung, H.S.H., Huang, R.Z., Liu, O.: Optimizing discounted cash flows in project scheduling - an ant colony optimization approach. IEEE Trans. Syst. Man Cybern. Part C **40**(1), 64–77 (2010)
8. Cohen, D., Gómez-Iglesias, A., Thiruvady, D., Ernst, A.T.: Resource constrained job scheduling with parallel constraint-based ACO. In: Wagner, M., Li, X., Hendtlass, T. (eds.) ACALCI 2017. LNCS (LNAI), vol. 10142, pp. 266–278. Springer, Cham (2017). https://doi.org/10.1007/978-3-319-51691-2_23
9. Dagum, L., Menon, R.: OpenMP: an industry-standard API for shared-memory programming. IEEE Comput. Sci. Eng. **5**(1), 46–55 (1998)
10. Demeulemeester, E., Herroelen, W.: Project Scheduling: A Research Handbook. Kluwer, Boston (2002)
11. Dorigo, M.: Optimization, learning and natural algorithms. Ph.D. thesis, Dip. Elettronica (1992)
12. Dorigo, M., Stützle, T.: Ant Colony Optimization. MIT Press, Cambridge (2004)
13. Gu, H., Schutt, A., Stuckey, P.J.: A lagrangian relaxation based forward-backward improvement heuristic for maximising the net present value of resource-constrained

projects. In: Gomes, C., Sellmann, M. (eds.) CPAIOR 2013. LNCS, vol. 7874, pp. 340–346. Springer, Heidelberg (2013). https://doi.org/10.1007/978-3-642-38171-3_24

14. Kimms, A.: Maximizing the net present value of a project under resource constraints using a lagrangian relaxation based heuristic with tight upper bounds. Ann. Oper. Res. **102**, 221–236 (2001)

15. Kolisch, R., Sprecher, A.: PSPLIB - a project scheduling problem library: OR software - ORSEP operations research software exchange program. Eur. J. Oper. Res. **96**(1), 205–216 (1997)

16. Merkle, D., Middendorf, M., Schmeck, H.: Ant colony optimization for resource-constrained project scheduling. IEEE Trans. Evol. Comput. **6**(4), 893–900 (2000)

17. Neumann, K., Schwindt, C., Zimmermann, J.: Project Scheduling with Time Windows and Scarce Resources: Temporal and Resource-Constrained Project Scheduling with Regular and Nonregular Objective Functions, vol. 508. Springer, Heidelberg (2003). https://doi.org/10.1007/978-3-540-24800-2

18. Show, Y.Y.: Ant colony algorithm for scheduling resource constrained projects with discounted cash flows. In: Proceedings of the Fifth International Conference on Machine Learning and Cybernetics, Dalain, China, pp. 176–180. IEEE (2006)

19. Thiruvady, D., Ernst, A.T., Singh, G.: Parallel ant colony optimization for resource constrained job scheduling. Ann. Oper. Res. **242**(2), 355–372 (2016)

20. Thiruvady, D., Singh, G., Ernst, A.T.: Hybrids of integer programming and ACO for resource constrained job scheduling. In: Blesa, M.J., Blum, C., Voß, S. (eds.) HM 2014. LNCS, vol. 8457, pp. 130–144. Springer, Cham (2014). https://doi.org/10.1007/978-3-319-07644-7_10

21. Thiruvady, D., Singh, G., Ernst, A.T., Meyer, B.: Constraint-based ACO for a shared resource constrained scheduling problem. Int. J. Prod. Econ. **141**(1), 230–242 (2012)

22. Thiruvady, D., Wallace, M., Gu, H., Schutt, A.: A lagrangian relaxation and ACO hybrid for resource constrained project scheduling with discounted cash flows. J. Heuristics **20**(6), 643–676 (2014)

23. Vanhoucke, M.: A scatter search heuristic for maximising the net present value of a resource-constrained project with fixed activity cash flows. Int. J. Prod. Res. **48**(7), 1983–2001 (2010)

24. Vanhoucke, M., Demeulemeester, E., Herroelen, W.: On maximizing the net present value of a project under renewable resource constraints. Manag. Sci. **47**(8), 1113–1121 (2001)

An Efficient Heuristic to the Traveling Salesperson Problem with Hotel Selection

Marques Moreira Sousa[1,2(✉)], Luiz Satoru Ochi[2],
and Simone de Lima Martins[2]

[1] Federal Institute of São Paulo, Campos do Jordão-SP 12460-000, Brazil
marques.sousa@ifsp.edu.br
[2] Computing Institute, Fluminense Federal University, Niterói-RJ 24210-346, Brazil
{satoru,simone}@ic.uff.br

Abstract. Traveling salesperson problem with hotel selection consists of determining a tour for the salesperson who needs to visit a predefined number of customers at different locations by taking into consideration that each working day is limited by time. If the time limit is accomplished, the salesperson must select a hotel from the set of available ones to spend the night. The aim is to minimize the number of necessary days to visit all customers spending the shortest possible travel time. We present an adaptive efficient heuristic based on the Iterated Local Search metaheuristic to solve available instances. The proposed heuristic is able to find good solutions for almost all instances and, in some cases, it is able to improve the quality of the best results found in literature, decreasing the number of trips necessary or time to travel along a tour. Moreover, the heuristic is fast enough to be applied to real problems that require fast responses.

Keywords: Optimization · Metaheuristic · Iterated Local Search
Traveling Salesperson Problem with Hotel Selection

1 Introduction

The current paper concerns a variant of the classical Traveling Salesperson Problem (TSP). A set of cities needs to be visited exactly once by salesperson in TSP. Therefore, the aim of the TSP is to find the best tour to minimize the use of resources such as time, distance or costs. The tour should start in a specific city and return to it after all cities are visited.

Many studies have dealt with this problem in the last decades and very good results were found for the available test instances [1]. New variants have emerged to explore different and more complex applications in the real world. A recent variant is the Traveling Salesperson Problem with Hotel Selection (TSPHS) [28], which can be defined as a hierarchical multi-period TSP variation that has the salesperson's working day limited by a maximum time.

© Springer Nature Switzerland AG 2019
M. J. Blesa Aguilera et al. (Eds.): HM 2019, LNCS 11299, pp. 31–45, 2019.
https://doi.org/10.1007/978-3-030-05983-5_3

Two terms are used to describe the difference between a single working day and the whole work accomplished by the salesperson, namely: trip, which is used to indicate a sequence of visited customers beginning at one hotel and ending in the same or another hotel (which can be an extreme or an intermediate hotel); tour, which means the set of connected trips that together attend all customers. The aim of the TSPHS is to lexicographically minimize the number of working days and the total travel time.

Formally, TSPHS takes into consideration the graph $G = (V, A)$; where: $V = H \cup C$ (H represents the non-empty set of available hotels and C is the set of customers who should be visited just once). The edges are defined by $A = \{(i, j) | i, j \in V, i \neq j\}$ where each edge (i, j) represents a connection between two customers, two hotels or a customer and a hotel. There is an associated visitation time τ_i to each customer $i \in C$ and $\tau_i = 0$, for all $i \in H$. The necessary time c_{ij} to travel from facility i to j is known for all facility combinations. The tour must start and end at the same specific hotel ($i = 0, i \in H$) and such hotel could be used as an intermediate hotel between two connected trips. The remaining hotels ($i \neq 0, i \in H$) can be used, if needed, in a complete salesperson tour. A hotel can be used more than once due to its features, and that is why a TSPHS solution cannot be expressed by a simple circle. Moreover, each trip should start and end at one of the available hotels. The total visitation time spent in one trip cannot exceed a predefined constant L; besides, one trip must start where the previous one ended. A formulation is proposed in [7].

Applications in the real world can be mapped within this problem: (i) a salesperson who needs to deliver many products but faces constraints associated with his/her luggage capacity, and who needs to pick-up products at some place; (ii) electrical vehicles that need to stop in specific recharge places; (iii) a truck driver who has predefined appropriate areas to stay in at night throughout the tour; and (iv) other applications that require partitioning the service required to be done.

The first study concerning TSPHS was proposed by Vansteenwegen et al. [28], who developed a two-index formulation and solved it using an approach based on the Iterated Local Search (ILS) heuristic [17]. Castro et al. [7] proposed a formulation based on the sub-tour Dantzig-Fulkerson-Johnson elimination constraints and, additionally, an effective approach consisting of a Memetic Algorithm (MA) [3] was adopted. A Tabu Search (TS) was applied to improve quality solutions. Although MA was able to obtain good solutions to all available TSPHS instances, it demands too much computational time for instances with more than 288 customers, which makes it unsuitable for real-time applications that require computational time not larger than a few seconds. A comparison between different memetic algorithm approaches was conducted by Sousa et al. [25]. Castro et al. [6], in order to decrease the computational time, have introduced a fast heuristic that requires very short computational time and finds solutions of good quality. Their heuristic uses the Iterated Local Search framework with the Variable Neighborhood Descent (VND) [13] procedure to replace the conventional local search procedure. Additionally, two perturbation procedures capable of bringing diversification to the solutions are iteratively applied.

Recently, some authors proposed and investigated TSPHS variants such as the Multiple Traveling Salesperson Problem with Hotel Selection (MTSPHS) [4] and the Traveling Salesman Problem with Multiple Time Windows and Hotel Selection (TSPMTW-HS) [2]. Other problems with similar features involving hotel selection (intermediate facilities) are: Orienteering Problem with Hotel Selection (OPHS) [11], Capacitated Arc Routing Problem with Intermediate Facilities [24], Vehicle-Routing Problem with Intermediate Replenishment Facilities [27] and Vehicle Routing Problem with Intermediate Facilities [18]. In fact, heuristic and metaheuristics such as Variable Neighborhood Descent, Tabu Search, Greedy Randomized Adaptive Search Procedure, Iterated Local Search and Memetic Algorithm have been used to effectively solve this kind of problem. Population heuristic often leads to solutions of better quality; however, they demand much more computational time. The key question for solving TSPHS is to find balance between quality and computational time.

Given the complexity of the classical Traveling Salesperson Problem and by considering that TSPHS can be easily turned into TSP just by eliminating the time limit constraint trip, TSPHS is at least as hard as TSP and, consequently, it is also $\mathcal{NP}$-hard. Thus, optimally solving moderate to big sized instances can be impractical, even with the use of currently available high-performance machines. The contribution of the present study lies on the development of an effective adaptive heuristic for TSPHS able to lead to solutions of good quality using little computational time.

Despite the similarities with other TSP variants, TSPHS has one feature that makes it harder to solve. The choice of intermediate hotels causes great impact on the final solution [14]. A simple change of an intermediate hotel has direct impact on at least two trips, thus it forces these trips to be re-optimized.

The current paper is organized as follows: Sect. 2 describes the herein proposed heuristic; Sect. 3 presents results and; finally, Sect. 4 relates conclusions, future research about TSPHS and the suggested variants.

2   A Heuristic for TSPHS

This section describes the proposed heuristic algorithm, which is based on ILS metaheuristic whose main steps are summarized in Algorithm 1. ILS was used to solve the related problem due to its simple structure, few parameters to set and high performance when applied to similar vehicle routing problems [8,23].

Algorithm 1. Procedure Iterated Local Search (ILS)

1: **out:** s^*
2: s^0 = GenerateInitialSolution()
3: s^* = LocalSearch(s^0)
4: **while** termination condition not met **do**
5:     s' = Perturbation(s^*);
6:     s'^* = LocalSearch(s');
7:     s^* = AcceptanceCriterion(s^*, s'^*);
8: **end while**

Initially, a feasible solution is generated and a local search is applied to the solution. Then, some iterations are performed until a termination criterion is met. A perturbation is applied to the solution to generate a new solution in each iteration. A local search is conducted in order to look for a better solution in the neighborhood of the modified solution. After the local search, in case the new generated solution passes an acceptance test, it becomes the next incumbent solution, otherwise the incumbent solution is not changed.

Algorithm 2 shows the developed Efficient Adaptive Iterated Local Search (EA-ILS) Procedure. An initial solution is built and improved through the local search VND [12] procedure (lines 2–3). Subsequently, a percentage parameter used by the perturbation procedure is defined (line 6) (such parameter will be explained later). The procedure demands performing some iterations and it is over when i_{max} iterations are performed without solution improvement (lines 7–24). A new solution from a neighborhood (choose randomly a hotel used in a solution and put then in another position inside the tour, too randomly defined) or from the perturbation procedure (see Subsect. 2.3) is obtained in each iteration (lines 8–12). The local search procedure is an attempt (line 13) to improve the solution. There is a variable that controls how often the perturbation procedure is applied (lines 21–23). The generation of a new solution through a perturbation or by choosing a neighbor always occurs considering the best current solution of a given iteration.

The diversification procedure is controlled and it allows increasing perturbation when the difficulty of improving a solution increases. If an iteration of the proposed heuristic procedure is multiple of the *var_mod* parameter, the perturbation procedure is applied. Otherwise, one neighbor solution of the incumbent solution is chosen to be explored. The interval established to apply the perturbation is defined through the *interval_pert* parameter with a value empirically defined. This procedure saves computational time and ensures that a neighborhood of a search space will be more explored before to move to another region.

The proposed heuristic uses a random mechanism to guide the solution through different search spaces; therefore, an efficient way to generate random numbers presented by [20] is used. This random number generating mechanism assures effectiveness and uncommonly generates repeated numbers.

Throughout the application of Algorithm 2, the feasibility is guaranteed only for the initial solution. Next, all generated solutions may be feasible or not. Thus, the solution found after the perturbation procedure is not necessarily feasible. However, such occurrence is desirable in order to enable diversification. The local search procedure uses a solution generated by the perturbation procedure and needs a way to check the quality of the solution in each iteration. Therefore the heuristic adopts an objective function (see Eq. (1) below) that penalizes unfeasible trips (contained inside a solution y) in order to analyze unfeasible solutions. Trips exceeding the time limit constraint have the infeasibility quantities multiplied by the constant $M = 10000$. Such M value was defined by Castro et al. [7]. Such penalty value forces the heuristic to give preference to better tours focusing in feasible solutions with smallest number of trips. In other words, this

Algorithm 2. Algorithm for finding a high quality TSPHS solution from a subset of all possible solutions

```
 1: in: i_max, interval_pert     out: y
 2: y = GenerateInitialSolution()
 3: y = LocalSearch(y)
 4: i = 0
 5: var_mod = interval_pert
 6: θ_1 = Define-Perturbation()
 7: while (i < i_max) do
 8:    if ((i % var_mod)≠ 0) then
 9:       y' = Neigborhood(y)
10:    else
11:       y' = Perturbation(y, θ_1)
12:    end if
13:    y' = LocalSearch(y')
14:    if y' better than y then
15:       y = y'
16:       i = 0
17:       var_mod = interval_pert + 1
18:    else
19:       i = i + 1
20:    end if
21:    if ((i%var_mod)= 0) then
22:       var_mod = var_mod − 1
23:    end if
24: end while
```

procedure ensures that a feasible solution with at least the same number of trips will always be better than another infeasible solution.

$$F(y) = \sum_{d=1}^{D}(M + time_d) + M\sum_{d=1}^{D}\max(0, time_d - L) \qquad (1)$$

The proposed heuristic uses the same neighborhood structures from Castro et al. [6], which had been applied to similar routing problems before obtaining satisfactory results. In this work, these neighborhoods have been used in a different context from other works because they were used for developing a Random Variable Neighborhood Descent (RVND) heuristic for (local search (see Subsect. 3.2).

Also a new way of applying the perturbation procedure was developed. There is a periodicity for applying the perturbation procedure in the current solution and the control of perturbation intensity is adjusted proportionally to instance size. Both are explained in Subsect. 2.3.

2.1   Initial Solution

The order-first split-second method was the procedure chosen to generate an initial solution, similarly to the method used by Castro et al. [6]. Initially, a TSP tour is generated using Lin-Kernighan heuristic (LK) [16] as implemented by [1]. The use of LK is necessary to avoid the initial solution to have a big number of trips and consequently, demands high computational processing time to achieve feasibility. For some complex instances, without using a LK it is impossible to find a feasible initial solution in reasonable computational time. Furthermore, this technique was widely used with success by [6,7].

Using LK, there is no time limit imposed to the trips, the tour starts and ends at a predefined hotel (hotel 0) and all the customers are visited. The initial solution obtained through LK for almost all cases is not feasible to TSPHS; thus, a split procedure is applied in a second phase in order to make each trip duration less than the time limit. This splitting procedure was inspired in Dijkstra Algorithm [10]. A graph is developed using all customers and hotels. All edges that do not comprise the LK solution are multiplied by the constant $M = 10000$ and it implies that Dijkstra's algorithm will tend to keep the generated solution LK. The Dijkstra procedure starts by visiting the initial hotel ($h = 0$) and it iteratively visits a customer as an attempt to form a minimum time path. When no more customers can be added to the path, without violating the time limit for a trip, then one of the hotels is chosen to be the trip stop. The final path involves customers and hotels that guarantee the viability of the solution, as well as that all customers are visited.

2.2   Local Search Procedure

The local search procedure is based on the simple Variable Neighborhood Descent (VND) [21]. Local search procedures exploring many neighborhood structures have been used to solve problems that demand tour definition. The procedure differs from the conventional VND because the application order of different neighborhood structures is randomly chosen in each iteration of Algorithm 2. The variation called RVND was effectively used in other optimization problems [19, 26]. Algorithm 3 describes the RVND procedure. Initially, the application order of the neighborhoods is randomly defined (line 2). Then, a local search is performed according to the order defined by index k.

If a better solution is found through the kth neighborhood, the variable k is set to 1 ($k = 1$) and the first neighborhood is used again. If, at any iteration, the local search does not return a better solution, the next neighborhood is used. The procedure ends when there is no better solution after all neighborhoods are tested.

An embedded search procedure (Algorithm 4) is applied in each RVND iteration. Neighborhood structures used are explained below in details.

- **2-OPT** (intra-trip) [9] makes a move inside a trip so that two distinct edges connecting customers are removed and the visiting order of the intermediary customers is reversed.

Algorithm 3. LocalSearch Algorithm

1: **in:** y **out:** y
2: DF[ ] = Define_Order()
3: k = 1
4: **while** (k <= 4) **do**
5:     $y' = \text{SEARCH}(N_{DF[k]}, y)$
6:     **if** (y' better than y) **then**
7:         $y \leftarrow y'$
8:         $k \leftarrow 1$
9:     **else**
10:         k = k + 1
11:     **end if**
12: **end while**

- **OR-OPT** (intra-trip) [22] reorganizes a set of m customers inside a trip; where: $m = \{3, 2, 1\}$. An inverse order $\{1, 2, 3\}$ is valid, however, is computationally more expensive (identical to Relocate and Exchange). This structure finds the neighbor solution that mostly improves the total travel time.
- **Relocate** (inter-trip) [15] modifies a solution moving a set of m customers from one trip to another. This neighborhood tries to move consecutive customers in a given order $m = \{3, 2, 1\}$ to find the neighbor with the best improvement by considering all tested neighbors.
- **Exchange** (inter-trip) [15] alters two different trips by exchanging a set of m consecutive customers; where: $m = \{3, 2, 1\}$, by considering the best solution to execute the movement, only.
- **Changehotels** (hotel operator) [6] defines a structure to test if the change of a hotel by another improves the incumbent solution. The quality of the final solution strongly depends on the choice made for the intermediate hotels [2, 7, 28].
- **Jointrips** (hotel operator) [6,7] tries to decrease the number of used intermediate hotels. Consequently, it joins two trips that share the same hotel. This operator works only with hotels. Without this operator, an improvement on the number of trips would not be possible; thus, the same number of trips defined on initial solution would be kept. This operator is quite relevant when considering the first TSPHS goal, that is to minimize the number of trips.

The procedure labeled as SEARCH receives two parameters: the type of neighborhood to be used and a solution to be improved. If the obtained solution y'' is better than y', then, two local searches using 2-OPT and Or-OPT neighborhood structures are applied as an attempt to improve the solution. The best neighbor y' is returned by the procedure SEARCH. The search procedure uses four neighborhood structures, which were successfully applied by [6,7,25,28]. They consist of two different approaches. The first one, which is called intra-trips structure, is applied inside a single trip. The second one is used to improve a solution by taking into consideration all trips that together make up the tour.

Algorithm 4. SEARCH Algorithm

1: **in:** N_k , y **out:** y'
2: $y' = y$
3: **while** (improvement is found) **do**
4:    $y'' = argmin_{S \in N_k(y')} F(S)$
5:    **if** (y'' better than y') **then**
6:       $y' = $ optimize y'' **with** 2-OPT/OR-OPT
7:    **end if**
8: **end while**

2.3 Perturbation

The perturbation method used to assure solution diversification consists of randomly choosing some customers of the current solution and moving them to other positions which are also randomly chosen. The first issue concerns defining a perturbation rate able to determine the number of customers to be moved. This number should be carefully defined in order to be large enough to guide the solution to another search space and, at same time, to be small enough to keep some of the features detected in the current solution. If these assumptions are not taken into account, the resultant solution may be too similar to the current solution or too different.

A fixed perturbation rate has been used in many studies [6,7]. However, when it comes to the TSPHS problem, problems can arise when a fixed rate is used. If the instance has a small number of customers, e.g. 10 customers, at least a rate of 10% is needed to perturb a single customer. However, if the same perturbation rate is applied to one of the biggest instances available to TSPHS, e.g. 1001 customers, 101 customers will be moved within the solution. The perturbation procedure applied to a larger number of customers may be successfully accomplished. However, the new solution may be too much different and worse than the present solution, thus leading the heuristic to a solution space too distinct from the one that has been exploited so far.

A varying rate was defined through the Nonlinear Symmetrical Sigmoidal Function (4PL) described by MyCurveFit (http://mycurvefit.com/) in order to prevent this problem. This function assures that a sufficient number of customers would be changed for a small instance. The changes would guide the search for a new search space without performing a completely random restart of the procedure. It is importante to know that a higher perturbation rate (20%) is applied to small instances (e.g. 10 customers), whereas this percentage is drastically decreased to 1% for large instances (see Fig. 1).

3 Computational Experiments

All computational experiments were carried out in an Intel i7 870 computer, with a 2.93 GHz processor, 8 GB RAM under Ubuntu 14.04 operational system and gcc 4.8.2 compiler using optimization flag -Ofast. The proposed heuristic

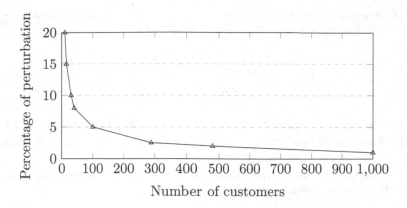

Fig. 1. Relationship between applied perturbation rate and number of customers.

used two parameters: the maximum number of iterations without improvement (i_{max}) and the interval of iterations without perturbation (*interval_pert*). The benchmark instances used in this paper were defined by Castro et al. [5].

Three values were tested for each parameter: 20, 30 and 40 for i_{max} and 3, 5 and 7 for *interval_pert*. Tests were conducted with the 9 possible combinations. The configuration that obtained the highest number of best results in relation to the best solution in the literature, spending the least computational time was 30 for i_{max} and 5 for *interval_pert*.

Perturbation strategy used by [6] considers a fixed percentage independent of instances characteristics (i.e. number of customers). The proposed heuristic (EA-ILS) innovates making use of an adaptive percentage based on the number of customers. An experiment was conducted to show effectiveness of this technique. The EA-ILS heuristic was executed using the adaptive rate percentage proposed in this work and the fixed rate percentage used by [6]. Each approach was executed 30 times, using all instances available. Results are summarized in Tables 1 and 2. In the first column Best Known Solution (BKS) is Best solution available so far and Perturbation Local Search (PLS) represents the results obtained by [6]. The next three columns present the number of trips (#T), average computational time (Time), and the average gap between the BKS and the value found by the heuristics. The gap value is calculated by Eq. (2). Results obtained in SET_2 with 10 and 15 customers were omitted since that there is no difference between the results obtained by BKS, PLS and the proposed heuristic. The number of trips found by the EA-ILS heuristic using adapted and fixed percentage rate for perturbation was the same. The gaps obtained by the adaptive strategy was equal or less than the gaps obtained by the fixed strategy for all groups of instances except group SET_3_5, while the computational times needed for the adaptive strategy is equal or less to those needed by the fixed strategy in all groups.

Once that EA-ILS with adaptive perturbation procedure reached good results, another test was conducted to check if all neighborhood structures are

Table 1. Summarized results comparing number of trips, computational time and GAP for Best Known Solution, PLS and EA-ILS, part 1.

	SET_1			SET_2_30			SET_2_40		
	#T	Time	GAP	#T	Time	GAP	#T	Time	GAP
BKS	77	96.6	-	25	0.0	-	31	0.0	-
PLS	78	0.5	0.34	26	0.0	1.02	33	0.0	2.98
EA-ILS adaptive perturbation	**77**	**1.0**	**0.25**	**25**	**0.0**	**1.30**	**32**	**0.0**	**2.62**
EA-ILS fixed perturbation	77	1.3	0.34	**25**	**0.0**	**1.30**	**32**	**0.0**	**2.62**

Table 2. Summarized results comparing number of trips, computational time and GAP for Best Known Solution, PLS and EA-ILS, part 2.

	SET_3_3			SET_3_5			SET_3_10			SET_4		
	#T	Time	GAP	#T	Time	GAP	#T	Time	GAP	#T	Time	GAP
BKS	65	60.6	-	97	44.2	-	169	42.0	-	92	81.1	-
PLS	67	2.5	0.31	98	1.6	0.38	173	1.0	−0.06	95	2.5	0.59
EA-ILS adaptive	**64**	**5.3**	−0.02	96	3.4	−0.15	**169**	**3.1**	−0.13	**91**	**33.7**	0.67
EA-ILS fixed	64	19.3	0.01	**96**	**18.2**	−0.16	169	17.3	−0.04	91	46.9	0.92

Table 3. Summarized results showing difference between execution of full neighborhood proposed heuristic and execution without each neighborhood structure separately, part 1.

	SET_1			SET_2_30			SET_2_40		
	#T	Time	GAP	#T	Time	GAP	#T	Time	GAP
EA-ILS	**77**	**1.0**	**0.25**	25	0.0	1.30	32	0.0	2.62
Without Relocate	79	0.2	0.66	25	0.0	1.30	33	0.0	2.93
Without Exchange	78	0.8	0.22	25	0.0	1.30	32	0.0	2.62
Without JoinTrips	79	1.1	0.23	27	0.0	1.02	34	0.0	2.62
Without Chghotels	77	1.2	0.28	25	0.0	1.34	32	0.0	2.65
Without 2OPT	77	1.3	0.26	25	0.0	1.30	32	0.0	2.64
Without OrOpt	77	1.2	0.26	25	0.0	1.30	32	0.0	2.62

really necessary. The experiments consisted in disabling neighborhoods one at a time. Results are presented at Tables 3 and 4, showing the better results in dark gray color. Results for SET_2 with 10 and 15 customers were omitted since the results were the same for all neighborhoods. The gaps values show that it is not possible to determine which neighborhood has the greatest influence in the quality of the solution related to total trip time, because the better results produce variation according with the type of neighborhood that is not used. However the EA-ILS heuristic with adaptive strategy was able to find the least number of trips for all groups of instances. Therefore this justifies the need to use all structures of neighborhoods combined with adaptive perturbation.

Table 4. Summarized results showing difference between execution of full neighborhood proposed heuristic and execution without each neighborhood structure, part 2.

	SET_3_3			SET_3_5			SET_3_10			SET_4		
	#T	Time	GAP	#T	Time	GAP	#T	Time	GAP	#T	Time	GAP
EA-ILS	**64**	**5.3**	**−0.02**	96	3.4	−0.15	169	3.1	−0.13	**91**	**33.7**	**0.67**
Without Relocate	65	1.3	0.06	97	0.9	−0.07	172	0.8	−0.33	99	2.3	4.54
Without Exchange	64	3.9	0.00	**96**	**2.4**	**−0.15**	169	2.2	−0.02	92	41.2	0.93
Without JoinTrips	66	6.1	0.04	97	3.8	−0.07	172	3.1	−0.34	101	34.5	0.19
Without Chghotels	**64**	**5.3**	**−0.02**	96	3.4	−0.15	169	3.1	−0.05	91	34.6	0.68
Without 2OPT	64	5.8	0.01	96	3.9	−0.12	**169**	**2.6**	**−0.13**	91	13.1	0.84
Without OrOpt	64	4.3	0.01	97	3.8	−0.10	169	3.2	−0.04	92	34.4	0.74

Solutions found by Castro et al. [7] are often the best solutions in literature, but their computational time is too large; therefore, only a quality comparison was made. The main focus of the present study consists in comparing the proposed heuristic and the fastest heuristic for the TSPHS that presented good solutions [6]. Therefore, just the computational times of PLS [6] and proposed Efficient Adaptive Iterated Local Search (EA-ILS) were compared. Since the machine used in the present study was similar to that used by Castro et al. [6, 7], it was not necessary to apply any rescaled method such as the one applied by Castro et al. [7].

The solution s_1 is considered to be better than solution s_2 in the presented results, if the number of trips in s_1 is smaller than the number in s_2, even if the total travel time in s_1 is larger than in s_2. Furthermore, if two solutions have the same number of trips, the one with the shorter total travel time is considered to be better.

Each instance was executed 30 times and the best result for all iterations is presented. The obtained results were compared by firstly considering the number of trips; next, the total travel time; and finally, by considering the computational time.

The Gain Average Percentage (GAP) equation used to compare the results obtained through the heuristic and the BKS only takes the total travel time into consideration, and it includes the visitation time to each customer. The reason to not consider the number of trips lies on the fact that all procedures led to the same number of trips in almost all instances. Positive GAP values mean that the solution is worse than BKS and negative values mean a solution is better than BKS. Some special cases occur when the GAP is positive but the number of trips is smaller because it makes the result better than BKS.

$$GAP = 100 \times \frac{Time(Heuristic) - Time(BKS)}{Time(BKS)}. \tag{2}$$

Due to the space limitation, the tables with detailing results found for each of the available instances were made available in an appendix format that can be accessed by the following url: https://goo.gl/5UfhBz.

The obtained results are presented summarized in Tables 5 and 6. In first table, are showed the average of results obtained. In this table first column shows the sets of instances used while second column information about average computational time spent on BKS. In next four columns, are demonstrated computational time and GAP for heuristic PLS and EA-ILS.

Table 5. Summary of results of all available sets of instances

	BKS	PLS		EA-ILS	
	Time	Time	GAP	Time	GAP
SET_1	96.6	0.5	0.34	1.0	0.25
SET_2_10	0.0	0.0	0.00	0.0	0.00
SET_2_15	0.0	0.0	0.00	0.0	0.00
SET_2_30	0.0	0.0	1.02	0.0	1.30
SET_2_40	0.0	0.0	2.96	0.0	2.62
SET_3_3	60.6	2.5	0.31	5.3	−0.02
SET_3_5	44.2	1.6	0.38	3.4	−0.15
SET_3_10	42.0	1.0	−0.06	3.1	−0.13
SET_4	81.1	2.5	0.59	33.7	0.67

For SET_1 is possible to see that EA-ILS GAP is approximately 26% smaller than that observed in PLS and the number of trips used is equal to BKS (see Table 6). Computational time is 0.5 s on average higher than PLS, however is an acceptable difference for application in real world problems. For SET_2_10 (10 customers) and SET_2_15 (15 customers) there are no difference significant both in computational time and in GAP. In SET_2_30 and SET_2_40 computational time is no expressive, meanwhile GAP values are higher for both heuristics, by the fact that BKS consider an exact approach with good results in SET_2. Nevertheless EA-ILS require fewer number of trips than PLS. For SET_3 there are three subsets of instances, with three, five and ten extra hotels. Results for all three subsets show superiority of EA-ILS in terms of quality solution becoming better than the BKS for some instances. Computational time for EA-ILS never exceeds three times the computational time spent by PLS. In reference to the last set, the GAP are quite higher than PLS, however does not mean worsening in the solution for all instances, since the number of trips required in EA-ILS is lower than that found by PLS. An important point is the computational time demanded for this group, which was much higher due to an excessive time consumed by the last and bigger instance. In general, EA-ILS stands out PLS on average quality solution, using a bit of more computational time.

According to the problem objective that says a solution with smaller number of trips is better, Table 6 presents the sum of trips for each set of instances. The results show that for all sets (except SET_2 with 40 customers) the sum of trips of EA-ILS is equal or smaller than BKS and PLS. This demonstrates the robustness of the method to find good solutions.

Table 6. Sum of trips needed for each set of instance for compared heuristics or heuristics and exact methods

	SET_1	SET_2				SET_3			SET_4
		Number of customers				Number of hotels			
		10	15	30	40	3	5	10	
BKS	77	14	16	25	31	65	97	169	92
PLS	78	14	16	26	33	67	98	173	95
EA-ILS	**77**	**14**	**16**	**25**	32	**64**	**96**	**169**	**91**

To show the effectiveness of the proposed approach, a statistical test was done in order to compare the solutions found by EA-ILS and solutions reported by Castro et al. [6]. A single solution for each instance was presented in Castro et al. [6], so to apply complex statistical tests or construct graphics (i.e. boxplot) to visualize variance intervals of solutions was impracticable.

Considering the first objective of TSPHS that is to minimize number of trips necessary to visit all customers, a statistical T-test was carried out comparing number of trips of PLS and EA-ILS. Two hypotheses were defined, being: (i) H_0 null hypothesis means that the difference between average number of trips is equal to zero; (ii) H_1 alternative hypothesis that there is difference between average number of trips, considering an alpha value of 0.01 (i.e. $\alpha = 0.01$). The tests conducted on Microsoft Excel 2013 accuse a *p-value* = 0.00002, so rejecting the null hypothesis.

4 Conclusions

The TSPHS is a challenging combinatorial optimization problem and it can model other problems that demand routing and choice of intermediate facilities.

TSPHS in the present study was solved using an ILS heuristic. The computational experiments showed that the Adaptive Iterated Local Search heuristic found results of very good quality in comparison to the Best Known Solutions. It was able to improve 6 total travel time values and decrease the number of trips in 4 instances using small computational times. The maximum mean GAP between the proposed heuristic approach and other heuristic available in the literature is 0.67%, which is low considering that the best number of trips in some instances have decreased and it implies in larger total travel time.

When compared to those presented by Castro et al. [6] (PLS), they showed that when applied to 131 available instances, the proposed heuristic was able to find equivalent or better quality solutions for 123 instances. For 16 instances EA-ILS decreased number of trips necessary. The statistical T-test shows that the results are statistically significant when compared with PLS, showing that the proposed heuristic is efficient for decreasing the number of trips.

These good results showed that the proposal of developing a RVND heuristic, using known neighborhood structures, for the local search of an ILS heuristic worked very well. Also, the use of proposed variable perturbation has assured that the perturbation avoids a randomly restart in any instance of the problem, and has also allowed getting away from local optimal solutions.

TSPHS extensions with other variants, including time windows, multiple salesperson with heterogeneous load constraints and hotel costs, can be considered in future studies. These proposed variants of the problem will help modeling more realistic real-life situations.

Acknowledgments. This study was financed in part by the Coordenação de Aperfeiçoamento de Pessoal de Nível Superior - Brasil (CAPES) - Finance Code 001. The authors gratefully acknowledge the financial support from Conselho Nacional de Desenvolvimento Científico e Tecnológico (CNPq), Federal Institute of São Paulo Campus Campos do Jordão (IFSP) and Computational Intelligence laboratory at Fluminense Federal University (LABIC) for supporting the development of this work.

References

1. Applegate, D.L., Bixby, R.E., Chvatal, V., Cook, W.J.: The Traveling Salesman Problem: A Computational Study. Princeton University Press, Princeton (2006)
2. Baltz, A., Ouali, M.E., Jäger, G., Sauerland, V., Srivastav, A.: Exact and heuristic algorithms for the travelling salesman problem with multiple time windows and hotel selection. J. Oper. Res. Soc. **66**, 615–626 (2014)
3. Berretta, R., Cotta, C., Moscato, P.: Memetic Algorithm. Wiley, Hoboken (2011)
4. Castro, M., Sorensen, K., Goos, P., Vansteenwegen, P.: The multiple travelling salesperson problem with hotel selection. Technical report, University of Antwerp, Faculty of Applied Economics, BE (2014)
5. Castro, M., Sörensen, K., Vansteenwegen, P., Goos, P.: A simple GRASP+VND for the travelling salesperson problem with hotel selection. Technical report, University of Antwerp, Faculty of Applied Economics, BE (2012)
6. Castro, M., Sörensen, K., Vansteenwegen, P., Goos, P.: A fast metaheuristic for the travelling salesperson problem with hotel selection. 4OR, pp. 1–20 (2014)
7. Castro, M., Sörensen, K., Vansteenwegen, P., Goos, P.: A memetic algorithm for the travelling salesperson problem with hotel selection. Comput. Oper. Res. **40**(7), 1716–1728 (2013)
8. Coelho, V.N., Grasas, A., Ramalhinho, H., Coelho, I.M., Souza, M.J.F., Cruz, R.C.: An ILS-based algorithm to solve a large-scale real heterogeneous fleet VRP with multi-trips and docking constraints. Eur. J. Oper. Res. **250**(2), 367–376 (2016)
9. Croes, G.A.: A method for solving traveling-salesman problems. Oper. Res. **6**(6), 791–812 (1958)
10. Dijkstra, E.W.: A note on two problems in connexion with graphs. Numerische mathematik **1**(1), 269–271 (1959)
11. Divsalar, A., Vansteenwegen, P., Cattrysse, D.: A variable neighborhood search method for the orienteering problem with hotel selection. Int. J. Prod. Econ. **145**(1), 150–160 (2013)

12. Hansen, P., Mladenović, N.: Variable neighborhood search: principles and applications. Eur. J. Oper. Res. **130**(3), 449–467 (2001)
13. Hansen, P., Mladenović, N., Brimberg, J., Pérez, J.A.M.: Handbook of Metaheuristics: Variable Neighbourhood Search. Springer, Heidelberg (2010)
14. Kim, B.I., Kim, S., Sahoo, S.: Waste collection vehicle routing problem with time windows. Comput. Oper. Res. **33**(12), 3624–3642 (2006)
15. Laporte, G., Gendreau, M., Potvin, J.Y.: Semet: classical and modern heuristics for the vehicle routing problem. Int. Trans. Oper. Res. **7**(4–5), 285–300 (2000)
16. Lin, S., Kernighan, B.W.: An effective heuristic algorithm for the traveling-salesman problem. Oper. Res. **21**(2), 498–516 (1973)
17. Lourenço, H.R., Martin, O.C., Stützle, T.: Iterated local search: framework and applications. In: Gendreau, M., Potvin, J.Y. (eds.) Handbook of Metaheuristics. ISOR, vol. 146, pp. 363–397. Springer, Boston (2010). https://doi.org/10.1007/978-1-4419-1665-5_12
18. Markov, I., Varone, S., Bierlaire, M.: Integrating a heterogeneous fixed fleet and a flexible assignment of destination depots in the waste collection VRP with intermediate facilities. Transp. Res. Part B: Methodol. **84**, 256–273 (2016)
19. Martins, I.C., Pinheiro, R.G., Protti, F., Ochi, L.S.: A hybrid iterated local search and variable neighborhood descent heuristic applied to the cell formation problem. Expert Syst. Appl. **42**(22), 8947–8955 (2015)
20. Matsumoto, M., Nishimura, T.: Mersenne twister: a 623-dimensionally equidistributed uniform pseudo-random number generator. ACM Trans. Model. Comput. Simul. **8**(1), 3–30 (1998)
21. Mladenović, N., Hansen, P.: Variable neighborhood search. Comput. Oper. Res. **24**(11), 1097–1100 (1997)
22. Or, I.: Traveling salesman-type combinatorial problems and their relation to the logistics of regional blood banking. Ph.D. thesis, Northwestern University, Evanston, Illinois (1976)
23. Penna, P.H.V., Subramanian, A., Ochi, L.S.: An iterated local search heuristic for the heterogeneous fleet vehicle routing problem. J. Heuristics **19**(2), 201–232 (2013)
24. Polacek, M., Doerner, K.F., Hartl, R.F., Maniezzo, V.: A variable neighborhood search for the capacitated arc routing problem with intermediate facilities. J. Heuristics **14**(5), 405–423 (2008)
25. Sousa, M.M., Gonçalves, L.B.: Comparação de abordagens heurísticas baseadas em algoritmo memético para o problema do caixeiro viajante com seleção de hotéis. In: Proc. XLVI Simpósio Brasileiro de Pesquisa Operacional, pp. 1543–1554. Salvador, Brasil (2014)
26. Subramanian, A., Battarra, M.: An iterated local search algorithm for the travelling salesman problem with pickups and deliveries. J. Oper. Res. Soc. **64**(3), 402–409 (2013)
27. Tarantilis, C.D., Zachariadis, E.E., Kiranoudis, C.T.: A hybrid guided local search for the vehicle-routing problem with intermediate replenishment facilities. INFORMS J. Comput. **20**(1), 154–168 (2008)
28. Vansteenwegen, P., Souffriau, W., Sorensen, K.: The travelling salesperson problem with hotel selection. J. Oper. Res. Soc. **63**(2), 207–217 (2011)

Strategies for Iteratively Refining Layered Graph Models

Martin Riedler[1]($\boxtimes$), Mario Ruthmair[2], and Günther R. Raidl[1]

[1] Institute of Logic and Computation, TU Wien, Vienna, Austria
{riedler,raidl}@ac.tuwien.ac.at
[2] Department of Statistics and Operations Research, University of Vienna,
Vienna, Austria
mario.ruthmair@univie.ac.at

Abstract. We consider a framework for obtaining a sequence of converging primal and dual bounds based on mixed integer linear programming formulations on layered graphs. The proposed iterative algorithm avoids the typically rather large size of the full layered graph by approximating it incrementally. We focus in particular on this refinement step that extends the graph in each iteration. Novel path-based approaches are compared to existing variants from the literature. Experiments on two benchmark problems—the traveling salesman problem with time windows and the rooted distance-constrained minimum spanning tree problem—show the effectiveness of our new strategies. Moreover, we investigate the impact of a strong heuristic component within the algorithm, both for improving convergence speed and for improving the potential of an employed reduced cost fixing step.

Keywords: Iterative refinement · Layered graphs
Integer programming · Traveling salesman problem with time windows

1 Introduction

Layered graphs (LGs) are a well-known technique in mathematical programming to deal with specific constraints and restrictions in problems expressed on graphs. The basic idea is to construct an extended model that considers some problem dimension explicitly to make it easier to formulate certain constraints or even impose them implicitly. Picard and Queyranne [12] were among the first to consider such an approach. They modeled the time-dependent traveling salesman problem by introducing for each original node copies for all sequence positions at which it might be feasibly reached. Another typical application is related to distance restrictions in graphs. In such cases one can create node copies w.r.t. the feasible distances at which the original nodes can be reached. By omitting copies beyond the distance limit it is implicitly ensured that all paths in the

Supported by the Vienna Science and Technology Fund through project ICT15-014.

extended graph adhere to the limit. If the dimension among which the original graph is extended corresponds to time, resulting LGs are sometimes called time-expanded networks. Such approaches are frequently considered for scheduling problems in which time is discretized to obtain so-called time-indexed models. For further details on LGs and associated mixed integer linear programming (MILP) formulations see the extensive survey by Gouveia et al. [8].

The main advantage of LG formulations is that they provide a convenient modeling option while usually leading to strong linear programming (LP) bounds. In many cases the LG is even acyclic and allows pseudo-polynomial formulations. However, there is also an important drawback involved: LGs and the associated models are typically much larger than simpler formulations on the original input. Frequently, this leads to models which are computationally impractical for reasonable problem sizes. However, often it is the case that already a subgraph of the full LG would suffice to encode an optimal solution. Several researchers used this observation to construct iterative algorithms that successively approximate the full LG until an optimal solution is found. This is usually done by omitting node copies and redirecting arcs. Among the first were Wand and Regan [16] who consider LG formulations for a pickup and delivery problem with time windows. In particular, they propose a relaxed formulation and a heuristic component that considers a subset of the feasible solutions. Those two formulations are successively extended until their bounds match, proving optimality. Ruthmair and Raidl [15] suggested such an iterative approach for the rooted distance-constrained minimum spanning tree problem (DCMST). Another successful application of a similar algorithm was proposed by Dash et al. [6] for solving the traveling salesman problem with time windows (TSPTW). Their approach differs slightly from the former two as it refines the reduced LG only based on solving LP relaxations in a first stage. The final reduced LG is then used for solving an MILP in which the remaining infeasibilities are tackled by cutting planes. Further iterative refinement approaches in the network design area were considered by Macedo et al. [11], Boland et al. [3,4], and Clautiaux et al. [5].

Algorithms of this type that contain a component for obtaining heuristic solutions provide an eventually converging sequence of primal and dual bounds. Therefore, such an algorithm can also be terminated prematurely to obtain a high-quality primal solution together with a dual bound.

All previous works in this area have in common that they consider only a single strategy for extending the reduced LG in each iteration without evaluating alternatives. The employed techniques reach from rather simple approaches to more complex algorithms. In our previous work [13] we presented an extensive evaluation of different refinement techniques for a resource-constrained project scheduling problem (RCPSP). However, scheduling problems are somewhat special when it comes to the refinement step. In an aggregated set of time instants it is usually not clear what is the best/most promising option to reveal infeasibilities. Network design problems appear to be more accessible in this respect. An LG relaxation is typically obtained by redirecting arcs due to omitted layers which causes the arcs to no longer represent correct lengths. This makes it straightforward how an arc can be corrected to the full extent. Moreover, we can use the knowledge regarding the true length of the arcs to get further insight from intermediate solutions. This is a crucial part of the algorithm and evaluating promising alternatives seems to be worthwhile.

In the following we start by introducing the necessary terminology of LGs in terms of an example problem: the TSPTW. In particular, we discuss how reduced LGs can be obtained, whose associated MILP models provide either primal or dual bounds. Then, we propose a generic refinement algorithm including several enhancements. Afterwards, we explain the specific refinement strategies and evaluate them in our computational experiments. In addition to existing strategies from the literature, we suggest new ones that aim at extracting more information from intermediate solutions. In the computational study we evaluate the discussed refinement strategies on several benchmark sets for the TSPTW. To show how the strategies behave on a structurally different problem we also conduct experiments for the rooted DCMST.

2   Mathematical Formalization

In the following we describe the construction of an LG and an associated MILP model. The process is exemplified in terms of the TSPTW. Afterwards, we characterize reduced LGs that serve as basis for the refinement algorithm introduced in the next section.

Notational Remarks. For a graph $G = (V, A)$ and node subset $S \subseteq V$ let $\delta^+(S) = \{(i, j) \in A \mid i \in S, j \notin S\}$ be the set of outgoing and $\delta^-(S) = \{(j, i) \in A \mid i \in S, j \notin S\}$ be the set of incoming arcs. To simplify notation we omit the set braces for singletons S. Variable vectors are denoted in bold face. Solution vectors are indicated by a superscript "$*$".

2.1   Traveling Salesman Problem with Time Windows

The traveling salesman problem with time windows (TSPTW) is defined on a directed graph $G = (V, A)$ with node set $V = \{\alpha, 1, \ldots, n, \omega\}$, associated arc costs $c \colon A \to \mathbb{Z}_{\geq 0}$, and travel times $t \colon A \to \mathbb{Z}_{> 0}$. As in [6], we represent the depot by two distinct nodes α and ω in order to model a tour starting and ending at the depot as path. Each node $i \in V$ is associated with a time window $[r_i, d_i]$ with $r_i \leq d_i$. Service times for the nodes can be incorporated into the travel times and are therefore not considered separately. The goal is to find a least cost Hamiltonian path through V starting at α and ending at ω s.t. all nodes are visited within their time windows. Waiting at nodes is allowed in case of early arrival.

2.2 Layered Graph Model

We consider the layered digraph $G_L = (V_L, A_L)$. Initially, node set $V_L = \{i_l \mid i \in V, l \in [r_i, d_i]\}$ contains all node copies that are feasible w.r.t. the time windows. Thereby, node copy $i_l \in V_L$ at layer l represents original graph node $i \in V$ reached at time l. To get an abstraction for connecting the layered node copies we introduce function $\theta(i_l, j) := \max(r_j, l + t_{(i,j)})$ that provides the layer at which node j is reached when starting at node i at layer l. The obtained arc set is $A_L = \{(i_l, j_m) \mid i_l, j_m \in V_L, (i, j) \in A, \theta(i_l, j) = m\}$. To obtain a smaller graph we remove all unnecessary node copies—and their incident arcs—that cannot be reached from depot copy α_{r_α}[1]. For an example see Fig. 1 where node 3_1 is not reachable from α_0.

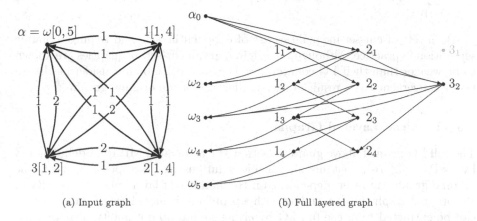

(a) Input graph   (b) Full layered graph

Fig. 1. Example of a layered graph for a TSPTW instance with $t = c$.

We model the TSPTW in terms of binary arc variables x_a, for $a \in A$, indicating which arcs of the original graph are part of the tour, and non-negative arc variables z_a for the LG. Observe that this problem could also be modeled without the original graph variables. However, these variables are convenient for imposing certain strengthening inequalities and beneficial for the reduced cost fixing explained in Sect. 3.

$$\text{(TSPTW-L)} \quad \min \quad \sum_{a \in A} c_a x_a \tag{1}$$

$$\text{s.t.} \quad \sum_{i_l \in V_L} \sum_{a \in \delta^-(i_l)} z_a = 1 \qquad \forall i \in V \setminus \{\alpha\}, \tag{2}$$

$$\sum_{a \in \delta^+(i_l)} z_a = \sum_{a \in \delta^-(i_l)} z_a \qquad \forall i_l \in V_L, \tag{3}$$

---

[1] When referring to the full LG G_L in the following, we assume this step to be completed.

$$\sum_{(i_l, j_m) \in A_L} z_{(i_l, j_m)} = x_{(i,j)} \qquad \forall (i,j) \in A, \qquad (4)$$

$$\boldsymbol{x} \in \{0,1\}^{|A|}, \; \boldsymbol{z} \in \mathbb{R}_{\geq 0}^{|A_L|}. \tag{5}$$

The model presented above can be strengthened[2] by well-known cut-set inequalities of the following form:

$$\sum_{a \in \delta^-(W)} x_a \geq 1 \qquad \forall W \subseteq V \setminus \{a\}, \; W \neq \emptyset. \tag{6}$$

Stronger cut-set inequalities can be specified w.r.t. the z variables (see [8]):

$$\sum_{a \in \delta^-(W)} z_a \geq 1 \qquad \forall W \subseteq V_L \setminus \{a_{r_\alpha}\}, \; W \neq \emptyset, \; \exists v \in V : \{v_l \in V_L\} \subseteq W. \tag{7}$$

Both sets of cut-set inequalities are of exponential size and require dynamic separation in practice. Although the original graph cut-set inequalities are known to be weaker than their LG counterpart, they are still worth considering due to faster convergence as a result of the smaller size of the original graph.

2.3　Reduced Layered Graphs

The full LG defined above guarantees that the associated MILP model, denoted by TSPTW-L(G_L), contains all feasible solutions that are possible w.r.t. the original graph. However, depending on the number of layers and the density of the original graph we often end up with a problematic model size. Smaller graphs can be extracted from the full LG by either giving up optimality or feasibility. For pragmatic reasons we require each reduced LG $G_L' = (V_L', A_L')$ to contain at least one copy for each node of the original graph, i.e., $V = \{i \mid i_l \in V_L'\}$.

Due to the omitted node copies, arcs are redirected. We say that an arc (i_l, j_m) is *shortened* (*lengthened*) if there exists an arc (i_l, j_k) in the full LG s.t. $m < k$ ($m > k$), otherwise it has the *correct* length.

Dual Layered Graphs. A *dual* LG $G_{dL} = (V_{dL}, A_{dL})$ is obtained by considering only a subset of the layered node copies $V_{dL} \subseteq V_L$ inducing the reduced arc set

$$A_{dL} = \{(i_l, j_m) \mid i_l, j_m \in V_{dL}, \, (i,j) \in A, \, m \leq \theta(i_l, j),$$
$$\nexists m' \, (m < m' \leq \theta(i_l, j) \wedge j_{m'} \in V_{dL})\}.$$

In short, this means that if a layered node is present in V_{dL} but the target of its outgoing arc according to G_L is not, then we use the copy of the target node at the maximum layer no larger than the originally used copy and omit the arc if such a copy does not exist. An example is provided in Fig. 2.

---

[2] In case of cycles due to zero travel times, these inequalities become mandatory.

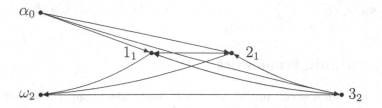

Fig. 2. Example of a dual layered graph w.r.t. the TSPTW instance provided in Fig. 1.

In order to guarantee that the associated MILP model is a relaxation we only consider node subsets $V_{dL} \subseteq V_L$ s.t. $i_l \in V_{dL} \wedge (i_l, j_m) \in A_L \implies \exists m' \ (m' \le m \wedge (i_l, j_{m'}) \in A_{dL})$. This ensures that we loose no connections that were present in the original graph. By using only shortened and correct arcs we never arrive at a node on a higher layer than in the full LG. As a result TSPTW-L(G_{dL}) is a relaxation w.r.t. the original problem. Consequently, the LP relaxation of TSPTW-L(G_{dL}) yields a dual bound. Moreover, if an optimal integral solution to TSPTW-L(G_{dL}) is feasible (on the x variables) w.r.t. the original problem, then it is guaranteed to be optimal. Observe that the dual LG is—opposed to the full LG—usually not acyclic. Therefore, separating cut-set inequalities is necessary to obtain a connected solution.

Primal Layered Graphs. A primal LG $G_{pL} = (V_{pL}, A_{pL})$ is obtained by considering only a subset of the layered node copies $V_{pL} \subseteq V_L$ and an associated induced arc set

$$A_{pL} = \{(i_l, j_m) \mid i_l, j_m \in V_{pL}, (i, j) \in A, m \ge \theta(i_l, j),$$
$$\nexists m' \ (\theta(i_l, j) \le m' < m \wedge j_{m'} \in V_{pL})\}.$$

This time we redirect arcs to the node copy at the minimum layer at least as large as the original one. Therefore, the primal LG turns its associated MILP model into a heuristic as it may exclude feasible solutions—possibly to the extent that no solutions remain. A feasible solution to TSPTW-L(G_{pL}) provides a primal bound but cannot be shown to be optimal on its own—not even if all layered arcs associated with the selected z variables have the correct length.

2.4   Other Problems

The definitions provided above can easily be adjusted to other problems. To cover the rooted DCMST (see [9])—for which we also perform experiments in Sect. 4—it suffices to redefine function θ to $\theta(i_l, j) := l + d_{(i,j)}$, i.e., waiting is not permitted/necessary. The problem's distance restriction can be imagined as time window for each node with a lower bound of zero and an upper bound equal to the global distance limit. A suitable MILP model can then be obtained by taking the model for the TSPTW and replacing constraints (3) by

$$z_{(i_l, j_m)} \leq \sum_{(k_h, i_l) \in \delta^-(i_l):k \neq j} z_{(k_h, i_l)} \quad \forall i_l \in V_L, \, \forall (i_l, j_m) \in \delta^+(i_l). \qquad (8)$$

3   Algorithmic Framework

In this section we describe our iterative refinement algorithm. In particular, we consider different refinement strategies that are used to iteratively extend an initially small dual LG.

To simplify the description in what follows, we make some assumptions on the considered input problem. We focus on problems for which each node must be connected to a designated source node. This guarantees that we can perform the necessary path computations in the LG for some of the refinement strategies presented in Sect. 3.2. Consequently, suitable connectivity inequalities must be available (e.g., cut-set inequalities (7) for the TSPTW). This assumption might seem restrictive at first but actually covers a large variety of problems. Depot based routing problems as well as most network design problems are compatible with this restriction. In addition, we assume that the model is specified as minimization problem and includes at least design variables x for the original graph arcs and design variables z for the LG variables—further auxiliary variables are of course possible. Note that these conditions are more strict than necessary but being more general would go beyond the scope of this work.

3.1   Iterative Refinement Algorithm

The main idea of our iterative refinement algorithm (IRA) is to start with a small dual graph and solve the associated MILP model or its corresponding LP relaxation, respectively. The result is then used to either prove optimality or—if this cannot be done—to obtain a larger dual graph (closer to the full LG) for repeating the procedure. This step of adding not-yet-present node copies of the full LG to the dual LG is called *refinement*. If the refinement adds at least one new node copy in each iteration, then it is guaranteed that the algorithm terminates with an optimal solution in finitely many iterations since the dual graph eventually converges to the full LG.

Algorithm 1 provides the detailed procedure. The mentioned gap refers to the absolute difference between the current dual (db) and primal (pb) bounds and is considered to be closed if $db \geq pb$. In the beginning we need an initial dual LG. This step depends on the problem at hand. For the TSPTW—and many other problems—a minimal starting graph that satisfies the restrictions imposed above can be obtained by considering for each original graph node the copy at the smallest feasible layer. Based on this initial dual LG we solve the LP relaxation of the associated MILP model. The obtained solution value is a dual bound and can be used to prove optimality if a primal bound is available. In order to get a more meaningful solution for the subsequent refinement process, we assume the LP to

---

Algorithm 1. Iterative refinement algorithm (IRA)

```
 1  while termination condition not met do
 2  |   solve LP; stop if gap closed
 3  |   refine LG
 4  |   if solution is integer and feasible then terminate     // optimal solution
 5  |   apply primal heuristic; stop if gap closed
 6  |   apply reduced cost fixing
 7  |   if graph could be refined then continue with next iteration
 8  |   solve IP
 9  |   refine LG
10  |   if solution is feasible then terminate                 // optimal solution
11  |   apply primal heuristic; stop if gap closed
12  end
```

---

be extended by connectivity inequalities. If optimality could not be proven yet, we use a refinement algorithm to identify possible infeasibilities in the relaxed solution. If infeasibilities could be detected, we add further nodes to the graph to reveal them. Otherwise, we test whether the obtained LP solution is integral. An integral solution that is feasible must be optimal according to the construction of the dual LG. A fractional solution, on the other hand, might prevent the refinement algorithm from detecting remaining infeasibilities. Therefore, we solve the MILP model in the following. However, before doing this, we can apply a heuristic (guided by the current fractional solution) to obtain a primal bound to possibly close the gap and prove optimality. Furthermore, we can use the obtained primal bound to attempt *reduced cost fixing* w.r.t. the x variables of the MILP model. To this end let db be the current solution value of the LP relaxation, pb the current primal bound, x^* the solution vector of the original graph variables, and x^r the vector of reduced costs of the x variables. For each arc $a \in A$ we consider two cases. If $x_a^* = 0 \wedge db + x_a^r \geq pb$, we can remove arc a from the input graph and consequently all its copies in any LG. On the other hand, if $x_a^* = 1 \wedge db - x_a^r \geq pb$, we know that arc a must be part of an optimal solution and its associated variable can therefore be fixed to one in subsequent iterations. When the algorithm is already close to convergence, reduced cost fixing might fix a sufficient number of variables to zero s.t. the model becomes infeasible, proving optimality of the solution that provided the current primal bound.

Unless one of the previous considerations allowed proving optimality, we are now in the situation that the (fractional) LP solution does not allow the refinement algorithm to identify the remaining infeasibilities. In this unfortunate case we have to take the additional computational burden and solve the MILP to optimality. If the integral solution is feasible, we proved optimality. Otherwise, we apply the primal heuristic once more before solving the LP relaxation according to the refined dual LG.

Primal Graph Heuristic. Steps 5 and 11 of Algorithm 1 can in principle be realized by any suitable heuristic. Problem-dependent algorithms typically provide better solution quality but are sometimes tedious to implement and often have to be replaced completely if a slightly different problem variant is considered. A more convenient problem-independent way to obtain heuristic solutions is to use the MILP formulation on the primal LG. We construct the primal LG by taking the node set of the dual LG and add for each node a copy at the maximum feasible layer. This enables us to benefit from the iterations made so far while reducing the risk of obtaining a graph that encodes no feasible solution. Primal graph heuristics of this type were considered in [5,14–16].

3.2   Refinement Procedures

The perhaps most crucial part of IRA is the refinement step. Solutions w.r.t. the current dual LG typically contain multiple infeasibilities and it is usually not clear how they can be handled most efficiently. In this context one has to deal (among others) with the following important questions: (a) for which nodes should further copies be added, (b) how many copies should be added, and (c) on which layers should the copies be inserted. Answering those questions typically involves keeping a suitable balance between the growth of the dual LG and the number of iterations IRA has to complete before proving optimality. The latter is quite important as it determines how often the MILP solver has to be invoked which is usually the most time-consuming part of the algorithm. On the other hand, the time each invocation takes increases with the size of the associated dual LG.

Full Infeasible Arc Refinement (FAR). The probably most straightforward refinement strategy simply refines all nodes that are part of the current solution. To this end, we consider all layered arcs whose associated solution value is non-zero. All shortened arcs are fully corrected by adding the appropriate target node to the dual LG. To avoid refining already feasible solutions, we check if the solution w.r.t. the x variables is feasible before starting the refinement process. This refinement procedure was employed in [4,11,14,15].

Infeasible Path Refinement (PR). Instead of considering all arcs, we only consider those that are most relevant from the structural perspective. We start by constructing an auxiliary LG that is obtained by taking all arcs of the dual LG with associated non-zero z variable value in the current solution. Based on this graph we compute a shortest path w.r.t. time, using travel times weighted by $1 - z_a^*$, to each node and determine the effective time at which the node would be reached. If the resulting arrival time is incompatible with the node's time window, we compute a refinement. This is done by traversing the path backwards and refining each arc as done for FAR stopping once we reach a node for which the effective time is equivalent to its layer.

Repeated Infeasible Path Refinement (RPR). We start by performing PR. In a subsequent step, we check for each formerly infeasible path if it is still contained in the adjusted dual LG—traversing different node copies but still reaching the target node after its time window when considering the effective length of the path—and repeat the refinement step until the path is no longer present.

Single-Copy Infeasible Path Refinement (SPR). Especially in later iterations the dual LG contains multiple layered copies w.r.t. each node of the original graph. We perform the same approach as done in PR, however, for each node of the original graph we only compute a refinement for the path reaching the node at the latest effective time, i.e., the most infeasible path.

Minimum Sum of Negative Waiting Times (DASH). This strategy was developed by Dash et al. [6]. Similar to the other techniques they consider the subgraph G'_{dL} induced by non-zero z variable values. If a node copy v_l is reached by shortened arcs, a new node copy is added that minimizes the sum of *negative waiting times*. The negative waiting time of an arc is essentially the amount by which it was shortened, i.e., if $(i_l, j_m) \in A'_{\mathrm{dL}}$ and $(i_l, j_k) \in A_{\mathrm{L}}$, then the negative waiting time is $k - m$. A new layered copy of node v is added at the layer λ that achieves the minimum when computing the sum of negative waiting times weighted by the associated arcs' solution values. Let z^* be the current solution vector and let $I_{v_l} = \{((j_m, v_l), l') \mid (j_m, v_l) \in A'_{\mathrm{dL}}, (j_m, v_{l'}) \in A_{\mathrm{L}}\}$. Then we seek the layer λ that minimizes $\mu(v_l, \lambda) = \sum_{(a,l') \in I_{v_l}: l' < \lambda} (l' - l) z^*_a + \sum_{(a,l') \in I_{v_l}: l' \geq \lambda} (l' - \lambda) z^*_a$. Dash et al. do not indicate which value is used if there are multiple options for λ that achieve the minimum. Function μ is piece-wise linear for a fixed v_l and changes slope only at points at which at least one arc arrives at the correct time. Therefore, it makes sense to restrict the procedure to such values, i.e., $l' - \lambda$ is zero for at least one element from I_{v_l}, as this guarantees to reduce the number of shortened arcs. However, this still might leave several options. In preliminary experiments we tested using either the smallest or the largest value of λ that achieves the minimum. The performance was roughly the same with a slight advantage for the latter.

Again we check feasibility w.r.t. the original graph variables to avoid superfluous refinements.

4 Computational Study

Our algorithms are implemented in C++ using CPLEX 12.8.0 as general-purpose MILP solver. All experiments have been performed in single thread mode with default parameter settings. For performance reasons the implicitly integral LG variables z are implemented as binary variables together with a cost-based branching priority to focus on the original graph variables. Experiments have been executed on an Intel Xeon E5540 machine with 2.53 GHz. The computation time limit has been set to 7200 s and the memory limit to 8 GB RAM. To test our framework we consider two benchmark sets for the TSPTW from http://lopez-ibanez.eu/tsptw-instances. The first set is from [1] and contains 50 instances while the second one was proposed in [7] and contains 135 instances. In addition, we also tested on instances for the rooted DCMST by [9]. Experiments were limited to the subset of 60 "TE" instances with distances ranging to 10, 100, and 1000 as the other instances turned out to be too easy. The TSPTW instances were preprocessed as described in [1] and the DCMST instances as described in [14].

We want to emphasize that our aim is to compare the different refinement strategies on a common basis and not to beat the state of the art. Achieving the latter would require further problem-specific tuning and incorporation of additional strengthening inequalities which is not the focus of this work.

4.1 Experiments

In the upcoming tables we present averages for gaps, computation times, the number of iterations in which the LP relaxation was solved (itr), the number of iterations in which the MILP was solved (itr-ip) as well as the number of nodes and arcs in the final LGs. Instances that terminated due to the memory limit are omitted when computing averages and those that ran into the time limit are considered with a value of 7200 s. Gaps are computed by $(pb^* - db)/pb^*$ where db is the dual bound of the respective run and pb^* is the best primal bound known for the respective instance. The remaining columns report the number of runs that ran into the time limit (tl) or the memory limit (ml), respectively, and the number of instances solved to proven optimality (opt).

For the TSPTW we consider each refinement strategy in three variants: without a primal component, with an initially provided primal solution ("HS"), and with an initially provided primal solution and reduced cost fixing activated ("HS_RCF"). We do this in order to show two things: (1) the benefits of a strong primal component and (2) the potential of reduced cost fixing if a high-quality solution is available. High-quality heuristic solutions for the Ascheuer instances were obtained from http://lopez-ibanez.eu/tsptw-instances and optimal solutions for the instances by Dumas et al. were obtained from http://homepages. dcc.ufmg.br/~rfsilva/tsptw.

Table 1 reports our results on the instances by Ascheuer et al. [1]. The first observation is that directly solving the MILP on the full LG (MIP) is not effective. The size of the associated model leads either to problems with the memory limit or to long runs that can frequently not be completed within the time limit. Consequently, the remaining gap is quite large with more than 10 % on average. All variants of our refinement algorithm perform much better. The most striking difference is that we deal with considerably smaller graphs that help to avoid any memory issues. This enables us to solve significantly more instances to optimality. Among the various refinement strategies we observe that the naive approach (FAR) solves the fewest instances to optimality while leading to the largest graph

Table 1. Results on the TSPTW instances by Ascheuer et al. [1]

Algorithm	Gap [%]	Time [s]	#itr	#itr-ip	$|V|$	$|A|$	#tl	#ml	#opt
MIP	20.24	3800	-	-	78437	1339490	21	5	24
IRA_FAR	0.09	2230	**17.9**	**0.0**	547	9820	14	0	36
IRA_DASH	0.08	2114	18.8	**0.0**	531	9546	13	0	37
IRA_PR	0.08	2090	31.4	3.5	**398**	**7663**	13	0	37
IRA_RPR	0.08	**2007**	22.1	2.5	417	8010	12	0	**38**
IRA_SPR	0.08	2080	32.6	3.6	399	7761	12	0	**38**
IRA_FAR_HS	0.08	2215	**15.1**	**0.0**	503	9248	14	0	36
IRA_DASH_HS	0.08	2064	15.5	**0.0**	480	8872	13	0	37
IRA_PR_HS	0.08	2038	27.9	3.2	**376**	**7379**	12	0	**38**
IRA_RPR_HS	0.08	**1937**	19.1	1.9	394	7710	12	0	**38**
IRA_SPR_HS	0.08	2068	29.2	3.1	377	7398	12	0	**38**
IRA_FAR_HS_RCF	0.08	2047	**15.3**	**0.0**	492	8340	13	0	37
IRA_DASH_HS_RCF	0.08	1935	15.4	**0.0**	475	7985	13	0	37
IRA_PR_HS_RCF	0.08	2096	29.2	3.6	**384**	**6876**	13	0	37
IRA_RPR_HS_RCF	0.08	**1933**	19.2	2.0	399	7190	12	0	**38**
IRA_SPR_HS_RCF	0.08	2001	30.6	3.3	384	6948	12	0	**38**

sizes. The approach by Dash et al. [6] works noticeably better and solves one more instance to optimality but requires comparatively large graphs. Our new path-based strategies solve the largest number of instances to optimality. The drawback of these rather careful and minimalist approaches is that they require a higher number of iterations to converge, even including some iterations where the MILP has to be solved, which is not necessary for FAR and DASH. Nevertheless, we observe the smallest average computation times for these strategies. The more slowly growing graphs outweigh the higher number of iterations through the smaller associated models. Among the three path-based approaches, we see that RPR performs best.

Providing high-quality initial solutions improves all variants of IRA alike. We observe a decrease in the number of iterations as well as the final graph sizes. This shows that a tight dual bound is sometimes obtained before feasibility can be established through further refinement steps. Enabling reduced cost fixing helps to improve the results further. For strategy PR we observe a minor slowdown compared to the variant in which only the initial solution is provided. The reason is that solution quality and refinement quality are not directly correlated. A weaker solution might lead to a very successful refinement in a subsequent iteration that is not reached by a better solution. The slowdown, however, is not dramatic and we achieve the smallest final graph size with this approach.

In Table 2 we provide the results on the instances by Dumas et al. [7]. Compared to the Ascheuer instances this set features much narrower time windows (100 at most). Therefore, the MILP on the full LG performs considerably better. Although it no longer faces problems with the memory limit, it is still not able

Table 2. Results on the TSPTW instances by Dumas et al. [7]

Algorithm	Gap [%]	Time [s]	#itr	#itr-ip	$|V|$	$|A|$	#tl	#ml	#opt
MIP	0.42	2000	-	-	3276	37069	29	0	106
IRA_FAR	0.00	250	10.5	0.2	285	3597	1	0	134
IRA_DASH	0.00	240	10.7	**0.1**	282	3549	1	0	134
IRA_PR	0.00	80	13.1	4.0	142	1719	0	0	**135**
IRA_RPR	0.00	**55**	**9.2**	3.0	145	1773	0	0	**135**
IRA_SPR	0.00	93	13.2	4.1	**141**	**1711**	0	0	**135**
IRA_FAR_HS	0.00	110	**8.2**	**0.1**	256	3207	0	0	**135**
IRA_DASH_HS	0.00	123	8.3	**0.1**	251	3135	0	0	**135**
IRA_PR_HS	0.00	65	12.3	3.5	139	1681	0	0	**135**
IRA_RPR_HS	0.00	**45**	8.4	2.5	139	1691	0	0	**135**
IRA_SPR_HS	0.00	73	12.4	3.5	**138**	**1674**	0	0	**135**
IRA_FAR_HS_RCF	0.00	44	**8.0**	**0.1**	255	2247	0	0	**135**
IRA_DASH_HS_RCF	0.00	45	8.3	**0.1**	250	2187	0	0	**135**
IRA_PR_HS_RCF	0.00	56	12.4	3.5	**138**	1395	0	0	**135**
IRA_RPR_HS_RCF	0.00	**42**	8.3	2.5	**138**	1401	0	0	**135**
IRA_SPR_HS_RCF	0.00	55	12.4	3.5	**138**	**1387**	0	0	**135**

to solve all instances to optimality within the time limit. The remaining gap is rather small but could also not be closed completely. In terms of computation times we again observe a clear advantage for IRA. The performance of the different refinement strategies is comparable to what we observed for the Ascheuer instances. Strategies FAR and DASH require larger graphs but converge within fewer iterations. The path-based approaches, on the other hand, lead to much smaller final graphs but also have to complete some iterations in which the MILP is solved. Providing an initial primal solution again improves the results significantly. This time reduced cost fixing provides a consistent improvement and does not suffer from side effects. It is even effective enough to almost improve the slower refinement strategies to the level of the better ones through variable fixes that significantly reduce the LG size.

Finding feasible solutions to the TSPTW is NP-hard (see [1]) but can be done in (pseudo-)polynomial time for the rooted DCMST. Therefore, we use this problem to show the performance of a simple problem-specific heuristic ("PHeu") in comparison to the general purpose heuristic based on the primal LG ("PG"). The considered problem-specific heuristic iteratively computes a resource-constrained shortest path to a still unreached node farthest from the source. The costs of the thereby added arcs are set to 0 for the next iteration and the procedure is stopped once all nodes are connected to the source. We use the solution on the x variables as guidance by operating on adjusted costs weighted by $1 - x_a^*$. We do not use reduced cost fixing here to avoid side effects that could influence the results. The outcome of the experiments on the DCMST instances is summarized in Table 3.

Table 3. Results on the hard DCMST instances (TE) by Gouveia et al. [9]

Algorithm	Gap [%]	Time [s]	#itr	#itr-ip	$|V|$	$|A|$	#tl	#ml	#opt
MIP	0.72	3063	–	–	23833	529314	13	13	34
IRA_FAR	0.08	2481	21.9	**0.0**	474	10827	9	0	51
IRA_DASH	0.05	2118	22.4	0.1	473	10816	6	0	54
IRA_PR	0.05	2555	23.6	0.2	418	9703	7	0	53
IRA_RPR	0.04	**1913**	**18.0**	0.1	481	10767	4	0	**56**
IRA_SPR	0.06	2601	24.3	0.2	**415**	**9640**	10	0	50
IRA_FAR_PHeu	0.07	2211	21.5	**0.0**	471	10769	6	0	54
IRA_DASH_PHeu	0.05	2058	21.9	**0.0**	472	10793	5	0	55
IRA_PR_PHeu	0.04	2256	23.0	0.1	417	9678	6	0	54
IRA_RPR_PHeu	0.04	**1680**	**17.8**	0.1	482	10781	4	0	**56**
IRA_SPR_PHeu	0.04	1987	23.6	0.2	**414**	**9632**	4	0	**56**
IRA_FAR_PG	0.08	2453	21.2	**0.0**	469	10709	9	0	51
IRA_DASH_PG	0.05	2185	21.5	**0.0**	467	10675	5	0	55
IRA_PR_PG	0.04	2269	23.3	0.2	416	9659	6	0	54
IRA_RPR_PG	0.04	**1816**	**17.6**	0.1	479	10702	4	0	**56**
IRA_SPR_PG	0.07	2553	23.9	0.2	**413**	**9606**	9	0	51

The MILP model on the full LG once more solves the fewest instances to optimality while being the slowest algorithm on average. The reason why the MILP is not that far off this time is that the considered instance set considers also small distance limits. For small and medium distance limits the MILP is competitive while it is clearly outperformed for the larger ones or cannot be solved due to the memory limit. Again, all variants of the iterative approach outperform the pure MILP approach for the larger distance restrictions. Strategies FAR and SPR do not work as well as the other strategies. The former appears to refine too unstructured while the latter does not make enough progress resulting in a comparatively high number of iterations. Among the remaining three variants we observe that RPR works best. Although it leads to larger graphs than the other path-based strategies, it turned out to be quite fast. Apparently, the repeated refinement helps to significantly reduce the number of iterations which compensates for the larger graph size.

Adding heuristics significantly decreases computation times and allows solving further instances to proven optimality. The primal LG heuristic turns out to be a valuable alternative to the problem-specific one. Nevertheless, we want to point out that it strongly depends on the problem whether the generic approach works well. Preliminary experiments for the TSPTW showed that it can be difficult to obtain feasible solutions if the underlying problem is challenging in this respect. Node copies corresponding to an initial heuristic solution might be inserted into the LG to resolve these issues.

Finally, we applied the one-tailed Wilcoxon signed-rank test for RPR and each other refinement strategy (without heuristics or reduced cost fixing). The alternative hypothesis that RPR is faster was assumed with a significance level of 0.05, except for the instances by Ascheuer where PR, SPR, and DASH performed too similar, mainly due to the comparatively high number of unsolved instances.

5   Conclusion and Future Work

In this work we considered a general framework for iteratively refining a reduced layered graph (LG). Based on solutions to an associated mixed integer linear programming formulation and heuristically obtained primal bounds the approach converges towards proven optimality if given enough time. In particular, we focused on one of the crucial points of such algorithms which did not receive much attention in previous works: the refinement step that extends the LG in each iteration. We investigated strategies from the literature and suggested new path-based ones. Through our experiments on two benchmark problems we could show that the previous approaches work reasonably well but still leave room for improvement. The path-based approaches are able to solve a higher number of instances to optimality while leading to smaller LGs in the final iteration. We also showed that a strong heuristic component is important for the algorithm to converge faster and to provide high-quality (intermediate) solutions.

A problem-independent heuristic was shown to be competitive with a simple problem-specific one. Future work could put more effort into this component to improve the obtained results. In this work we focused on refinement strategies for the reduced LG that is used to compute dual bounds. However, one may also consider refinements based on the LG that is used to obtain heuristic solutions.

For brevity, we restricted the discussion to problems with a designated source node to which all other nodes must be connected. However, the method can in principle be applied to any network design problem for which feasibility of the relaxation may be checked by path computations. Interesting problems are those whose resource dependencies can be naturally modeled through LG. This especially includes problems with resource-dependent (non-linear) costs, e.g., time-dependent travel times. If many layers are present of which only few are assumed to be traversed, the iterative algorithm is expected to work particularly well. In general, the approach does not work for applications represented by a cyclic LG because the described dual LGs not necessarily represent a relaxation for them. Typical examples are pickup and delivery problems with increasing and decreasing load along the route as well as the energy state in electric vehicle routing.

To be comparable to the state of the art further tuning would be necessary for both benchmark problems. In terms of the traveling salesman prblem with time windows our algorithms struggle in particular with some of the harder Ascheuer instances. A promising solution appears to be the inclusion of information related to node precedences as done, e.g., in [2,6]. Concerning the rooted distance-constrained minimum spanning tree problem our results are already quite close to the state-of-the-art column generation approach in [10] with only four instances that could not be solved to optimality.

In preliminary experiments we tested a cleanup algorithm that removes nodes from the LG that were not used for a specified number of iterations. This approach showed potential to decrease the final graph sizes further. Unfortunately, we ran into problems with cycling that increased the number of iterations, negating the provided benefits. Future research could address these issues by more complex cleanup or cycle-prevention strategies. Having shown that even smaller final graph sizes can be achieved, we think that a more theoretical investigation could prove useful. Computing minimal or even minimum LGs that lead to tight dual bounds or optimal solutions could serve as starting point to design more elaborate refinement strategies.

References

1. Ascheuer, N., Fischetti, M., Grötschel, M.: Solving the asymmetric travelling salesman problem with time windows by branch-and-cut. Math. Program. Ser. B **90**(3), 475–506 (2001)
2. Baldacci, R., Mingozzi, A., Roberti, R.: New state-space relaxations for solving the traveling salesman problem with time windows. INFORMS J. Comput. **24**(3), 356–371 (2012)
3. Boland, N., Hewitt, M., Marshall, L., Savelsbergh, M.: The continuous-time service network design problem. Oper. Res. **65**(5), 1303–1321 (2017)
4. Boland, N., Hewitt, M., Vu, D.M., Savelsbergh, M.: Solving the traveling salesman problem with time windows through dynamically generated time-expanded networks. In: Salvagnin, D., Lombardi, M. (eds.) CPAIOR 2017. LNCS, vol. 10335, pp. 254–262. Springer, Cham (2017). https://doi.org/10.1007/978-3-319-59776-8_21
5. Clautiaux, F., Hanafi, S., Macedo, R., Voge, M.E., Alves, C.: Iterative aggregation and disaggregation algorithm for pseudo-polynomial network flow models with side constraints. Eur. J. Oper. Res. **258**(2), 467–477 (2017)
6. Dash, S., Günlük, O., Lodi, A., Tramontani, A.: A time bucket formulation for the traveling salesman problem with time windows. INFORMS J. Comput. **24**(1), 132–147 (2012)
7. Dumas, Y., Desrosiers, J., Gelinas, E., Solomon, M.M.: An optimal algorithm for the traveling salesman problem with time windows. Oper. Res. **43**(2), 367–371 (1995)
8. Gouveia, L., Leitner, M., Ruthmair, M.: Layered graph approaches for combinatorial optimization problems. Comput. Oper. Res. **102**, 22–38 (2019)
9. Gouveia, L., Paias, A., Sharma, D.: Modeling and solving the rooted distance-constrained minimum spanning tree problem. Comput. Oper. Res. **35**(2), 600–613 (2008). Part Special Issue: Location Modeling Dedicated to the memory of Charles S. ReVelle
10. Leitner, M., Ruthmair, M., Raidl, G.R.: Stabilizing branch-and-price for constrained tree problems. Networks **61**(2), 150–170 (2013)
11. Macedo, R., Alves, C., de Carvalho, J.M.V., Clautiaux, F., Hanafi, S.: Solving the vehicle routing problem with time windows and multiple routes exactly using a pseudo-polynomial model. Eur. J. Oper. Res. **214**(3), 536–545 (2011)
12. Picard, J.C., Queyranne, M.: The time-dependent traveling salesman problem and its application to the tardiness problem in one-machine scheduling. Oper. Res. **26**(1), 86–110 (1978)

13. Riedler, M., Jatschka, T., Maschler, J., Raidl, G.R.: An iterative time-bucket refinement algorithm for a high-resolution resource-constrained project scheduling problem. Int. Trans. Oper. Res. (2017). https://doi.org/10.1111/itor.12445
14. Ruthmair, M.: On solving constrained tree problems and an adaptive layers framework. Ph.D. thesis, TU Wien, Vienna (2012)
15. Ruthmair, M., Raidl, G.R.: A layered graph model and an adaptive layers framework to solve delay-constrained minimum tree problems. In: Günlük, O., Woeginger, G.J. (eds.) IPCO 2011. LNCS, vol. 6655, pp. 376–388. Springer, Heidelberg (2011). https://doi.org/10.1007/978-3-642-20807-2_30
16. Wang, X., Regan, A.C.: Local truckload pickup and delivery with hard time window constraints. Transp. Res. B **36**(2), 97–112 (2002)

Fixed Set Search Applied to the Traveling Salesman Problem

Raka Jovanovic[1(✉)], Milan Tuba[2], and Stefan Voß[3,4]

[1] Qatar Environment and Energy Research Institute (QEERI),
Hamad Bin Khalifa University, PO Box 5825, Doha, Qatar
rjovanovic@hbku.edu.qa

[2] Department of Technical Sciences, State University of Novi Pazar,
Vuka Karadzica bb, 36300 Novi Pazar, Serbia

[3] Institute of Information Systems, University of Hamburg,
Von-Melle-Park 5, 20146 Hamburg, Germany

[4] Escuela de Ingenieria Industrial, Pontificia Universidad Católica de Valparaíso,
Valparaíso, Chile

Abstract. In this paper we present a new population based metaheuristic called the fixed set search (FSS). The proposed approach represents a method of adding a learning mechanism to the greedy randomized adaptive search procedure (GRASP). The basic concept of FSS is to avoid focusing on specific high quality solutions but on parts or elements that such solutions have. This is done through fixing a set of elements that exist in such solutions and dedicating computational effort to finding near optimal solutions for the underlying subproblem. The simplicity of implementing the proposed method is illustrated on the traveling salesman problem. Our computational experiments show that the FSS manages to find significantly better solutions than the GRASP it is based on, the dynamic convexized method and the ant colony optimization combined with a local search.

Keywords: Metaheuristic · Traveling salesman problem · GRASP

1 Introduction

In the last several decades there has been an extensive research effort on developing different metaheuristics for finding near optimal solutions for hard optimization problems. Most metaheuristic approaches focus on how to balance the global search (exploration) and local search (exploitation) in examining the solution space. There have been several directions in this research. Early methods include simulated annealing [23] and tabu search [15,16] where the search is focused near the best found solution and on mechanisms of escaping local optima. In later stages population based methods have proven to be very powerful. The general approach in such methods is generating a large number of solutions and including different types of learning mechanisms. In case of genetic algorithms

© Springer Nature Switzerland AG 2019
M. J. Blesa Aguilera et al. (Eds.): HM 2019, LNCS 11299, pp. 63–77, 2019.
https://doi.org/10.1007/978-3-030-05983-5_5

[27] and differential evolution [28] the main idea is in combining different high quality solutions with the addition of a certain level of randomization. Particle swarm optimization [1,2] explores the solution space through generating new solutions based on the positions of the globally and locally best found solution. This basic idea has been incorporated in a wide range of similar methods like cuckoo search [14], artificial bee colony algorithm [22] and many others. (The reader should note that we are fully aware about the controversial discussion with respect to some of these or similar approaches; see [31].) The ant colony optimization [10,20] uses a population based method to add a learning mechanism to greedy algorithms.

One of the most common methods for improving population based metaheuristics is by combining them with local searches. The variable neighborhood search [17] metaheuristic focuses on the efficient use of local searches. The performance of the original metaheuristics is often improved by different types of enhancements or by creating hybridized methods that combine one or more of such metaheuristic methods [5,25,34]. The main problems with such methods is the increased complexity of implementation. This problem is most evident if we observe publications in fields other than operations research and applied mathematics. In the vast majority of them only the original, simple to implement, method is used to solve the problem of interest.

Model-based heuristics are generally based upon the identification of a set of parameters, defining a model that, in turn, well captures some features of the search space [6]. These algorithms heavily rely on a set of update schemes used to progressively modify the model itself such that the possibility of obtaining higher quality solutions under the new model is increased. Recently, more and more emphasis is put on the application of learning mechanisms. In this phase, modifications are applied to the model and/or its parameters to reflect insights collected and generated during the search phase.

Well-known paradigms that can be interpreted under the philosophy of model-based heuristics are mostly from the area of *Swarm Intelligence*, but also focus on semi-greedy heuristics, including the greedy randomized adaptive search procedure (*GRASP*) [12,18], where the greedy function that guides the selection of the best candidates might incorporate some sort of stochastic model. Semi-greedy heuristics and GRASP exemplify of how simplicity is important. Although it generally has a worse performance than combining one of the more complex methods with a local search it is extensively used. The advantage of more complex metaheuristics often occurs only for very large problem instances; some examples in case of ACO can be seen in [19,21]. Because of this, it is reasonable to attempt to increase the size of problems that GRASP can solve, but in a way that there is no or only a small increase in complexity of the original method.

In this paper we focus on developing this type of method through adding a simple learning mechanism to GRASP. Some examples of such methods are GRASP with path relinking [13] and the dynamic convexized method [37]. Both of these methods produce a significant level of improvement. Note that both use the standard concept of intensifying the search of the solution space based on the location of globally and locally best found solutions. The basic concept of the proposed *fixed set search method* (FSS) is to avoid focusing on specific high quality solutions but on parts or elements that such solutions have. This idea of exploiting elements that belong to high quality solutions is used in ACO, were the randomized greedy algorithm is directed to choose such elements. The concept of generating new solutions based on the frequency of elements appearing in high quality solutions is the basis of the cross-entropy method (CE) [9]. The main conceptual difference between FSS and these two methods is that the proposed methods use only elements that are a part of locally optimal solutions. In practice this produces a significant difference in performance since the CE and ACO tend to over popularize elements in the best solution and in small variations of it and the FSS does not.

The ideas for developing this method may be based on earlier notions of chunking [35, 36], vocabulary building and consistent chains [30] as they have been used, e.g., in relation to tabu search. In those notions one relates given solutions of an optimization problem as composed of parts (or chunks). Considering the traveling salesman problem (TSP), for instance, a part may be a set of nodes to be visited consecutively. Moreover, some parts may be closely related to some other parts so that a corresponding connection can be made between two parts. Similar ideas are even found in the POPMUSIC paradigm [33]. The general idea of the proposed approach is to fix a set of elements that exist in high quality solutions and dedicate computational effort on "filling in the gaps". The idea of fixed sets has also been explored in the construct, merge, solve & adapt (CMSA) [4] matheuristic. In CMSA, which may also be interpreted as an implementation of POPMUSIC, a fixed set is used to decide which part of the solution space will be explored using an exact solver; later the newly generated solution is used to direct the next step of the search.

The concept of using fixed sets is illustrated on the symmetric TSP through adding a learning mechanism to GRASP. We should note that exact codes for the TSP are available [7]. Nevertheless, due to its widespread investigation it seems appropriate to use it for illustration purposes. As it will be seen in the following, this type of approach can easily be added to existing GRASP algorithms and produces a high level of improvement in the quality of found solutions and computational cost.

The paper is organized as follows. In the next section we give a brief description of GRASP for the TSP. Then we present the FSS and show how it is applied to the TSP. In Sect. 4, we discuss the performed computational experiments.

2   GRASP

In this section we provide a short outline of the GRASP used for solving the TSP. (A pseudocode for the general GRASP is given in Algorithm 1.) In the case of the TSP it is common to use a randomization of the nearest neighbor greedy algorithm with a restricted candidate list (RCL) based on the cardinality

---

Algorithm 1. Pseudocode for GRASP

---

while Not Stop Criteria Satisfied **do**
    Generate Solutions S using randomized greedy algorithm
    Apply local search to S
    Check if S is the new best
end while

---

of nodes [3]. For the local search, the most commonly used ones are the 2-OPT [8] and 3-OPT [24] searches. In practice instead of the original versions of the two local searches it is common to use a RCL of edges that will be used for evaluating the proposed improvement.

3   Fixed Set Search

In this section we present the proposed fixed set search metaheuristic and show how it can be used in combination with GRASP. Before giving the details of the method we give the basic concepts on which it is constructed.

One of the main disadvantages of GRASP is the fact that it does not incorporate any learning mechanism. On the other hand, such an improvement should be designed in a way that it is simple to implement. In this paper we propose one such method called the fixed set search (FSS). In the following we will assume that a solution S of the problem of interest can be represented in the form of a set. In case of the TSP, the solution S can be viewed as a set of edges $\{e_1, e_2, \ldots, e_l\}$. The development of FSS is based on two simple premises:

- A combinatorial optimization problem is generally substantially easier to solve if we fix some parts of the solution, and in this way lower the size of the solution space that is being explored.
- There are some parts of high quality solutions that are "easy to recognize". We say this in the sense that they appear in many good solutions. In general there is no need to dedicate a significant amount of computational effort to analyze them.

The general idea of FSS is to fix such "easy to recognize" parts of good solutions and dedicate computational effort in finding the optimal (or close to optimal) solution for the corresponding subset of the solution space. Informally, we take the common sections of good solutions, which we will call the *fixed set*, and try to "fill in the gaps". In the following sections we will illustrate how this simple idea can be incorporated in the GRASP metaheuristic for the TSP. The proposed algorithm has three basic steps. The first one is finding a fixed set. The second is adapting the randomized greedy algorithm to be able to use a preselected set of elements. Finally, specify and apply the method which gains experience from previously generated solutions.

3.1   Fixed Set

As previously stated, to be able to implement the proposed method it is necessary that we can represent a solution of the problem in the form of a set S. In case of the TSP, a solution S corresponds to the set of edges that represent a Hamiltonian cycle. Let us use the notation $\mathcal{P}$ for the set of all the generated solutions (population). In relation, let us define $\mathcal{P}_n$ as the set of n best generated solutions based on the objective function, the path length. Further, let us use the notation F for a *fixed set* that will be used in the search. Note that the elements of F will be inside the newly generated solution. In the following we define a method for finding a *fixed set* F for a population of solutions $\mathcal{P}$. The proposed method should satisfy the following requirements:

- (R1) A generated fixed set F should consist of elements of high quality solutions.
- (R2) The method should be able to generate many different random fixed sets that can be used to generate new high quality solutions.
- (R3) A generated fixed set F can be used to generate a feasible solution. More precisely, there exists a feasible solution S such that $F \subset S$
- (R4) Ability to control the size of the generated fixed set $|F|$.

The first two requirements can be achieved if we only use some randomly selected high quality solutions for generating the fixed sets. This can be achieved by simply selecting k random solutions from the set $\mathcal{P}_n$. Let us define $\mathcal{S}_{kn}$ as the set of selected solutions. The initial idea is to use the intersection of all the solutions in $\mathcal{S}_{kn}$ for the fixed set F. The problem is that we have no control over the size of the intersection. A simple idea to control the size of F is, instead of using the intersection of $\mathcal{S}_{kn}$, to select the elements (edges) that are part of the highest number of solutions. The problem with this approach is that such a selection can potentially contain edges that could not be used to generate a feasible solution.

Both of these issues can be avoided if a base solution $B \in \mathcal{P}_m$ is used in generating a fixed set F. More precisely, we can select the elements of B that occur most frequently in $\mathcal{S}_{kn}$. Let us define this procedure more formally. We will assume that we are finding a fixed set F with $|F| = Size$ for a set of solutions $\mathcal{S}_{kn} = \{S_1, .., S_k\}$ and base solution $B = \{e_1, \ldots e_l\}$. Let use define the function $C(e_x, S)$ which is equal to 1 if $e_x \in S$ and 0 otherwise. Using $C(e_x, S)$ we can define a function that counts the number of times an edge e_x occurs in $\mathcal{S}_{kn}$ as follows.

$$O(e_x, \mathcal{S}_{kn}) = \sum_{S \in \mathcal{S}_{kn}} C(e_x, S) \tag{1}$$

Now, we can define $F \subset B$ as the set of edges e_x that have the largest value of $O(e_x, \mathcal{S}_{kn})$. In relation, let us define function $F = Fix(B, \mathcal{S}_{kn}, Size)$ that corresponds to the fixed set generated for a base solution B, a set of solutions $\mathcal{S}_{kn}$ having $Size$ elements. An illustration of the method for generating a fixed set for the TSP can be seen in Fig. 1.

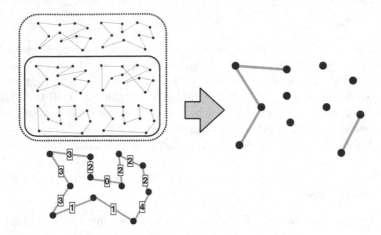

Fig. 1. Illustration of generating a fixed set. The input is $\mathcal{S}_{kn}$ (top left), a set of four randomly selected solutions out of the six best ones, and a base solution B (left bottom). Values on an edge of B represent the number of occurrences of that edge in elements of $\mathcal{S}_{kn}$. The edges on the right present the corresponding fixed set of size four.

3.2 Randomized Greedy Algorithm with Preselected Elements

To be able to use the fixed set within a GRASP setting we need to adapt the greedy randomized algorithm. Let us first note that in case of the TSP, the fixed set F will consist of several paths (sequence of edges which connect a sequence of vertices) of graph G. This effects the greedy algorithm in two ways. First, the inside nodes of the path should be removed from the candidate list. Secondly, if a node that is a start or end node of a path in the fixed set, is added to the current partial solutions the whole path must be added in the proper direction. Pseudocode for the adapted greedy algorithm can be seen in Algorithm 2.

In relation, let us define the function $S = RGF(F)$, for a fixed set F, as the solution acquired using this algorithm.

3.3 Learning Mechanism

In this section we present the FSS which is used as a learning mechanism for GRASP. Before presenting details of the proposed methods, let us first make a few observations. In the general case the early iterations of GRASP frequently manage to improve the quality of the best found solution. At later stages such improvements become significantly less frequent and the method becomes dependent on "lucky" hits. The idea is to use a fixed set F, for some promising region of the solution space, generate a solution $S = RGF(F)$ and apply a local search to S. In this way we increase the probability that a higher quality solution will be found. An important aspect is how to select the size of the fixed set. In case it is small, it efficiently performs a global search but after a certain number of executions, as in the case of GRASP, it will to a large extend be dependent on

---

Algorithm 2. Pseudocode for Greedy algorithm with preselected elements

---

Set *Paths* to all paths in F
Candidates $= V$
Candidates $=$ *Candidates* $\setminus F$
for all $p \in Paths$ **do**
 Candidates $=$ *Candidates* $\cup p[First] \cup p[Last]$
end for
Select Random start city from *Candidates*
while Not Completed Tour **do**
 Select next city c using RCL
 for all $p \in Paths$ **do**
  if $(c = p[First]) \wedge (c = p[Last])$ **then**
   Add path p to current solution in correct direction
   break
  end if
 end for
end while

"lucky" hits. On the other hand if F is large, it will only explore the parts of the solution space that are close to already generated solutions. As a consequence, there is a high risk of being trapped in locally optimal solutions.

This indicates that the size of the fixed set should be adapted during the execution of the algorithms. For simplicity, we can a priori define an array *Sizes* of fixed set sizes that will be tested, using the following formula:

$$Sizes[i] = |V| - \frac{|V|}{2^i} \qquad (2)$$

In (2), V represents the set of nodes of the graph on which the TSP is to be solved. The maximal value of an element in the array *Sizes* is chosen based on the problem being solved. Using this array let us give an outline of the FSS. We will first generate an initial population of solutions $\mathcal{P}$ by executing GRASP for N iterations. This initial population will be used to find the fixed sets. We start from a small fixed set and generate solutions until stagnation, in the sense of not finding new best solutions for a large number of iterations, has occurred. When stagnation occurs, we increase the size of the fixed set (selecting the next element of *Sizes*) and repeat the procedure. In this way a more focused exploration of the search space is executed. This procedure is repeated until the largest element in *Sizes* is tested.

At this stage it is expected that the set $\mathcal{P}_m$ of m best solutions has significantly changed and contains higher quality solutions than in the initial population. Because of this, there is a potential that even for smaller sized fixed sets, since they are now generated using better solutions, there is a higher probability of finding new quality solutions. So, we can repeat the same procedure from the smallest fixed set size. Let us note that after a large number of solutions is generated the new solutions acquired using small fixed sets are rarely new best ones. The importance of their revisit is in generating new types of high quality solutions. If the method does not manage to find a new solution among the best m ones for a large number of iterations for a specific fixed set size, this size can be excluded from the further search. This idea is better understood by observing the pseudocode for FSS given in Algorithm 3.

Algorithm 3. Pseudocode for the fixed set search

Initialize *Sizes*
$Size = Sizes.Next$
Generate initial population $\mathcal{P}$ using $GRASP(N)$
while (Not termination condition) **do**
    Set $\mathcal{S}_{kn}$ to random k elements of $\mathcal{P}_n$
    Set B to a random solution in $\mathcal{P}_m$
    $F = Fix(B, \mathcal{S}_{kn}, Size)$
    $S = RGF(F)$
    Apply local search to S
    $\mathcal{P} = \mathcal{P} \cup \{S\}$
    if Stagnant Best Solution **then**
        if (Stagnant Candidates) $\wedge$ $(Size = Min(Sizes))$ **then**
            Remove *Size* from *Sizes*
        end if
        $Size = Sizes.Next$
    end if
end while

In the pseudocode for the FSS, the first step is initializing the sizes of fixed sets using (2). Next the initial population of solutions is generated performing N iterations of the basic GRASP algorithm. The current size of the fixed set $Size$ is set to the smallest fixed set size. In the main loop, we first randomly generate a set of solutions $\mathcal{S}_{kn}$ by selecting k elements from $\mathcal{P}_n$. Next, we select a random solution B out of the set $\mathcal{P}_m$. Using $\mathcal{S}_{kn}$, B and $Size$ we generate the fixed set F as described in the above. Using F we generate a solution $S = RGF(F)$ using the randomized greedy algorithm with preselected elements. Next, we apply the local search to S and check if we have found a new best solution and add it to the set of generated solutions $\mathcal{P}$. After a new solution is generated we check the two stagnation conditions. The first one checks if the search for the best solution has become stagnant. If so, we set the value of $Size$ to the next value in $Sizes$. Let us note, that the next size is the next larger element of array $Sizes$. In case $Size$ is already the largest size, we select the smallest element in $Sizes$. Before updating $Size$, we also check if stagnation has occurred in the search of high quality solutions (we have not found a solution which is among the best n or m ones). In case this is true the current $Size$ is removed from $Sizes$. It is important to note, that this is only done if $Size$ is equal to the smallest member of $Sizes$. This is due to the fact that if we have managed to find an improvement for a smaller fixed set than $Size$, it is expected that we have just been "unlucky" and there is no need to remove this value from the search. This procedure is repeated until the array $Sizes$ is empty or some other termination criterion is satisfied.

4 Results

In this section we present the results of our computational experiments used to evaluate the performance of the proposed method. This has been done in a comparison with the GRASP algorithm presented in [26], the dynamic convexized method (DCTSP) from [37] and the ant colony optimization combined with a the 2-OPT local search (ACO-2OPT) [32]. The focus of the comparison is on the quality of found solutions.

The FSS and GRASP have been evaluated for both 2-OPT and 3-OPT as local searches. In case of the DCTSP a combination of 2-OPT and 3-OPT has been used as a local search. GRASP has been included to be able to evaluate the effect of the learning mechanism included in the FSS. DCTSP has been used in the comparison since it is a good representative of a metaheuristic whose search is focused on regions near the best solution. The comparison with ACO-2OPT is used to show the advantage of having a learning mechanism dependent on the local search, which is the case for FSS but not ACO-2OPT. In case of the FSS the randomized greedy algorithm used an RCL with 20 elements. In case of both local searches, 2-OPT and 3-OPT, the same size of RCL has been used. To increase the computational efficiency of the local searches we have used the standard approach of "don't look bits" (DLBs) [3]. Note that when we apply the local search inside the main loop of FSS, some of the DLBs could be preset based on the fixed set which significantly decreased the computational cost. The parameters for FSS are the following; $k = 10$ random solutions are selected from the best $n = 500$ ones for the set of solutions S_{kn}. The base solution is selected from the $m = 100$ best solutions. The size of the initial population was 100. The stagnation criterion was that no new best or high quality solution has been found in the last $Stag = 100$ iterations for the current fixed set size. The FSS and GRASP with 2-OPT have been implemented in C# using Microsoft Visual Studio 2017. The calculations have been done on a machine with Intel(R) Core(TM) i7-2630 QM CPU 2.00 GHz, 4 GB of DDR3-1333 RAM, running on Microsoft Windows 7 Home Premium 64-bit.

The comparison of the methods has been done on the standard benchmark library TSPLIB [29]. The test instances are the same as in [37]. A total of 48 test instances with Euclidean distances are used, with the number of nodes ranging from 51 to 2392. Note that in FSS the fact that distances are Euclidean is not exploited. The termination criterion was that a maximal number of solutions has been generated. The limit was the same as in [37], more precisely in case of problem instances having less than 1000 nodes it was $100|V|$, with V the set of nodes of the considered instance, and in case of larger instances it was $10|V|$. For each of the instances a single run of each of the methods has been performed, as in [37]. The results of the computational experiments can be seen in Table 1. In it, the results for DCTSP are taken from [37] and for GRASP with 3-OPT are taken from [26]. Note that the results for GRASP-3OPT are very similar to the ones from [26] and slightly better than our implementation. In case of $ACO - 2OPT$, the results for the same number of generated solutions and applied local searches have been taken from [32]. Note that these results correspond to an average of ten independent runs.

From the results in Table 1, we can first observe that the GRASP-2OPT has a significantly worse performance than all other methods finding best known solutions for only four instances and having an average relative error of 2.73%. In case of FSS-2OPT, the improvement is very significant: 20 known best solutions are found and the average relative error is 0.40%. What is very interesting is that FSS-2OPT performs only slightly worse than GRASP-3OPT which finds 22 known best solutions and has an average relative error of 0.39%. This indicates that the use of the proposed method can be very beneficial in case of less powerful local searches. The FSS-2OPT has found a higher quality solution than ACO-2OPT for each of the test instances from [32]. The average relative error of FSS-2OPT, on these instances, is 0.36% compared to 1.65% of ACO-2OPT.

Although FSS-2OPT has an overall worse performance than DCTSP, it manages to find better solutions for five problem instances. From the results in Table 1, it is evident that FSS-3OPT has the best performance. It manages to find better solutions than all the other methods for all instances or equal in case methods have found best known solutions. It finds three more known optimal solutions than DCTSP and has a notable improvement in relative average error. It is important to note that FSS-3OPT never has an error greater than 0.40%.

The parameters selected for specifying FSS have been chosen empirically through extensive testing. Overall the FSS is not highly sensitive to these parameters. The parameter k used to specify the number of solutions selected for generating the set S_{kn} had the following effect. In case of small values, the selection would result in highly randomized fixed sets. The reason for this is that there are no clear "good elements", especially in case of larger problem instances where there are not many common elements in all the solutions. In case of large values of k, the method would select very similar fixed sets. The parameter m used to specify the population from which the base solution would be selected has the following effect. In case of small values of m the convergence speed would initially be very fast but would quickly get trapped in locally optimal solutions. This is due to the fact that it becomes very hard to find new high quality solutions. Such values of m are useful in case we can only generate a small number of solutions. In case of high values of m, the convergence speed is much slower. The problem is that when the fixed set is generated for a lower quality solution B, although the method manages to find solutions of higher quality than B, it is unlikely that very high quality ones will be found. The effect of parameter n for specifying the size of the population used for generating S_{kn} had a similar effect but to a much lower extent. We found that the best choice of parameters for stagnation, both for finding best and high quality solutions, was the same as the number of iterations from which GRASP would rarely find new best solutions. In general, it is important to avoid very small values for this parameter because it results in prematurely stopping the evaluation of small fixed sets.

In Table 1 we did not include computational times, since they are highly dependent on structures used for implementing the local searches. The same

Table 1. Comparison of the proposed algorithms with GRASP and DCTSP for different TSPLIB instances.

Instance	Tour length							Relative error [%]					
	2OPT			3OPT			Known	2OPT			3OPT		
	GRASP	ACO	FSS	GRASP	DCTSP	FSS	best	GRASP	ACO	FSS	GRASP	DCTSP	FSS
eil51	426	-	426	426	426	426	426	0.00	-	0.00	0.00	0.00	0.00
berlin52	7542	-	7542	7542	7542	7542	7542	0.00	-	0.00	0.00	0.00	0.00
pr76	108351	-	108159	108159	108159	108159	108159	0.18	-	0.00	0.00	0.00	0.00
rat99	1223	-	1211	1211	1211	1211	1211	0.99	-	0.00	0.00	0.00	0.00
kroA100	21282	21427	21282	21282	21282	21282	21282	0.00	0.68	0.00	0.00	0.00	0.00
kroB100	22157	-	22141	22141	22141	22141	22141	0.07	-	0.00	0.00	0.00	0.00
kroC100	20802	-	20749	20749	20749	20749	20749	0.26	-	0.00	0.00	0.00	0.00
kroD100	21468	-	21309	21294	21294	21294	21294	0.82	-	0.07	0.00	0.00	0.00
kroE100	22106	-	22100	22068	22068	22068	22068	0.17	-	0.15	0.00	0.00	0.00
rd100	7960	-	7910	7910	7910	7910	7910	0.63	-	0.00	0.00	0.00	0.00
eil101	638	-	629	629	629	629	629	1.43	-	0.00	0.00	0.00	0.00
lin105	14379	-	14379	14379	14379	14379	14379	0.00	-	0.00	0.00	0.00	0.00
pr107	44394	-	44303	44303	44303	44303	44303	0.21	-	0.00	0.00	0.00	0.00
pr124	59159	-	59030	59030	59030	59030	59030	0.22	-	0.00	0.00	0.00	0.00
ch130	6135	-	6110	6110	6110	6110	6110	0.41	-	0.00	0.00	0.00	0.00
pr136	98614	-	96920	96772	96772	96772	96772	1.90	-	0.15	0.00	0.00	0.00
pr144	58554	-	58537	58537	58537	58537	58537	0.03	-	0.00	0.00	0.00	0.00
ch150	6586	-	6549	6528	6528	6528	6528	0.89	-	0.32	0.00	0.00	0.00
kroA150	26768	-	26524	26524	26525	26524	26524	0.92	-	0.00	0.00	0.00	0.00
pr152	74315	-	73682	73682	73682	73682	73682	0.86	-	0.00	0.00	0.00	0.00
rat195	2391	-	2330	2331	2323	2323	2323	2.93	-	0.30	0.34	0.00	0.00
d198	16000	15856	15803	15788	15780	15786	15780	1.40	0.48	0.15	0.05	0.04	0.00
kroA200	29803	-	29368	29380	29382	29368	29368	1.48	-	0.00	0.04	0.00	0.00
kroB200	29909	-	29447	29482	29437	29437	29437	1.60	-	0.03	0.15	0.00	0.00
ts225	127485	-	127301	126643	126643	126643	126643	0.66	-	0.52	0.00	0.00	0.00
pr226	80714	-	80369	80414	80369	80369	80369	0.43	-	0.00	0.06	0.00	0.00
gil262	2456	-	2378	2385	2379	2378	2378	3.28	-	0.00	0.29	0.04	0.00
pr264	50744	-	49135	49135	49135	49135	49135	3.27	-	0.00	0.00	0.00	0.00
a280	2658	-	2584	2589	2579	2579	2579	3.06	-	0.19	0.39	0.00	0.00
pr299	49522	-	48256	48235	48207	48191	48191	2.76	-	0.13	0.09	0.03	0.00
lin318	43324	42426	42185	42538	−	42029	42029	3.08	0.94	0.37	1.21	-	0.00
rd400	15986	-	15322	15385	15299	15284	15281	4.61	-	0.27	0.68	0.12	0.02
fl417	12066	-	11883	11895	11883	11871	11861	1.73	-	0.19	0.29	0.19	0.08
pr439	110564	-	107259	107401	107303	107217	107217	3.12	-	0.04	0.17	0.08	0.00
pcb442	52790	51794	50945	50946	50860	50846	50778	3.96	2.00	0.33	0.33	0.16	0.13
d493	36192	-	35055	35253	35136	35018	35002	3.40	-	0.15	0.72	0.38	0.05
att532	28965	28233	27860	28180	−	27735	27686	4.62	1.98	0.62	1.78	-	0.17
rat575	7143	-	6795	6863	6814	6776	6773	5.46	-	0.32	1.33	0.61	0.04
p654	35113	-	34812	34707	34658	34645	34643	1.36	-	0.49	0.18	0.04	0.01
d657	51226	-	49258	49531	49110	49014	48912	4.73	-	0.71	1.27	0.40	0.21
rat783	9352	9142	8869	8897	8848	8815	8806	6.20	3.81	0.72	1.03	0.48	0.10
pr1002	276251	-	264737	262060	260218	259512	259045	6.64	-	2.20	1.16	0.45	0.18
pcb1173	61210	-	57788	57676	57061	56965	56892	7.59	-	1.57	1.38	0.30	0.13
d1291	54537	-	51026	51616	51099	50862	50801	7.35	-	0.44	1.60	0.59	0.12
rl1304	270441	-	255867	255185	253842	253361	252948	6.92	-	1.15	0.88	0.35	0.16
rl1323	288538	-	271837	273115	271914	270678	270199	6.79	-	0.61	1.08	0.63	0.18
fl1400	21044	-	20398	20310	20167	20149	20127	4.56	-	1.35	0.91	0.20	0.11
fl1577	23274	-	22512	22427	22352	22300	22249	4.61	-	1.18	0.80	0.46	0.23
rl1889	339151	-	322883	319250	317825	317801	316536	7.14	-	2.01	0.86	0.41	0.40
d2103	86179	-	81197	81312	81078	80450	80551	7.12	-	0.93	1.07	0.78	0.13
pr2392	409970	-	387169	386017	380030	379307	378032	8.45	-	2.42	2.11	0.53	0.34
Number of found best known solutions (instances from [37])								2	-	20	22	27	31
Average relative error (instances from [37])								2.73	-	0.40	0.39	0.15	0.05

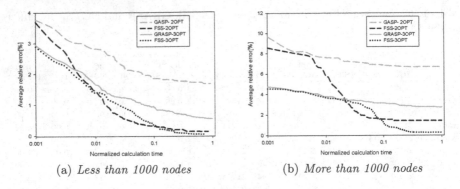

(a) *Less than 1000 nodes*   (b) *More than 1000 nodes*

Fig. 2. Average relative error for multiple TSPLIB problem instances with normalized execution time

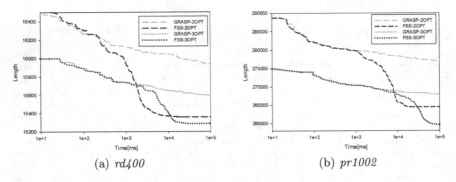

(a) *rd400*   (b) *pr1002*

Fig. 3. Average solution quality of ten independent runs of FSS and GRASP for the TSPLIB problem instances

information is excluded in the articles used for comparison. We would like to note that in case of our implementation of FSS and GRASP there was a significant decrease in computational time. This is due to the fact of a smaller candidate set for the greedy algorithm. The second reason is that the number of iterations needed to generate the solution was significantly lower. As it is well-known, the computational cost of 2-OPT and 3-OPT local searches is significantly higher than for generating the initial solution. An extensive analysis of the computational cost of 2-OPT can be found in [11]. In case of FSS, we exploit the fact that the fixed set is a subset of a locally optimal base solution B through DLBs. More precisely, the DLBs of all the inner points of paths in the fixed set can be preset. Note that the number of preset DLBs is close to the size of the fixed set. In practice this means instead of having the computational cost of the first iteration of 2-OPT being proportional to $|V|C$, were C is the size of the RCL, it is close to $(|V| - |F|)C$. Similar analysis can be done for the 3-OPT local search. In case of very large fixed sets, the time FSS generated a new solution and applied the local search was a fraction of the time needed to accomplishing the same task in GRASP. It is expected that a similar behavior would be present in applying FSS to other combinatorial problems.

This decrease in computational cost of the FSS compared to GRASP directly effects the convergence speed. In Fig. 2 we show the average relative error for instances having up to/above 1000 nodes for normalized time. The normalization has been done based on the maximal time needed for all the methods to find the best solutions for each instance. From these results it is evident that the use of FSS significantly increases the convergence speed. Another illustration of this behavior can be seen in Fig. 3 for representative problem instances. In each of the figures the convergence speed of the average solution length for ten independent runs of GRASP and FSS, with 2-OPT and 3-OPT used as local searches, are shown. It can be observed that there is a drastic increase in the convergence speed after the initial population is generated for the FSS.

5 Conclusion

In this paper we have presented a new metaheuristic called fixed set search that exploits the common elements of high quality solutions. The proposed meta-heuristic represents a method of adding a learning mechanism to the GRASP metaheuristic. It is expected that FSS can be applied to a wide range of problems since the only requirement is that the solution can be represented in a set form. A very important aspect of FSS is the simplicity in which a GRASP algorithm can be adapted to it. This is done with two basic steps. Firstly, the randomized greedy algorithm, used in the GRASP, is adapted to a setting were some elements are preselected. We have shown that this can be trivially achieved in case of the TSP, and it is expected that this is the case for many other combinatorial problems. Secondly, the method for generating a fixed set needs to be implemented which consists in selecting several solutions and tracking the number of times their elements occur in a selected base solution.

We have illustrated the effectiveness of the proposed approach on the TSP. Our computational experiments have shown that the proposed method has a significantly better performance than the basic GRASP approach when both solution quality and convergence speed are considered. Further, we have shown that the approach has a considerably better performance than the dynamic con-vexized method applied to the TSP in case 3-OPT is used as a local search. The proposed method has proven very efficient in improving the performance of GRASP in case of a less powerful local search.

It is important to note that there is a wide range of potential improvements to the proposed method. Some examples are having a more intelligent method of selecting the solutions used in generating the fixed set, or adapting the com-putational effort used to solve the subproblem related to a specific fixed set. Our objective was to show that even in the most basic form the proposed method can produce a significant improvement. We would like to note that the concept of using a fixed set can potentially be used to hybridize other metaheuristics like ACO, genetic algorithms and similar, for solving large scale problems through focusing the search in some promising areas of the solution space. On the other hand the idea of fixing elements of a solution can easily be included in mixed integer programs so there is a potential of adapting FSS to a matheuristic setting.

References

1. Banks, A., Vincent, J., Anyakoha, C.: A review of particle swarm optimization. Part I: background and development. Nat. Comput. **6**(4), 467–484 (2007)
2. Banks, A., Vincent, J., Anyakoha, C.: A review of particle swarm optimization. Part II: hybridisation, combinatorial, multicriteria and constrained optimization, and indicative applications. Nat. Comput. **7**(1), 109–124 (2008)
3. Bentley, J.J.: Fast algorithms for geometric traveling salesman problems. ORSA J. Comput. **4**(4), 387–411 (1992)
4. Blum, C., Pinacho, P., López-Ibáñez, M., Lozano, J.A.: Construct, merge, solve & adapt a new general algorithm for combinatorial optimization. Comput. Oper. Res. **68**, 75–88 (2016)
5. Blum, C., Puchinger, J., Raidl, G.R., Roli, A.: Hybrid metaheuristics in combinatorial optimization: a survey. Appl. Soft Comput. **11**(6), 4135–4151 (2011)
6. Caserta, M., Voß, S.: Metaheuristics: intelligent problem solving. In: Maniezzo, V., Stützle, T., Voß, S. (eds.) Matheuristics: Hybridizing Metaheuristics and Mathematical Programming, vol. 10, pp. 1–38. Springer, Boston (2010). https://doi.org/10.1007/978-1-4419-1306-7_1
7. Concorde: Concorde TSP solver (2015). http://www.math.uwaterloo.ca/tsp/concorde/index.html
8. Croes, G.A.: A method for solving traveling-salesman problems. Oper. Res. **6**(6), 791–812 (1958)
9. De Boer, P.T., Kroese, D.P., Mannor, S., Rubinstein, R.Y.: A tutorial on the cross-entropy method. Ann. Oper. Res. **134**(1), 19–67 (2005)
10. Dorigo, M., Blum, C.: Ant colony optimization theory: a survey. Theor. Comput. Sci. **344**(2–3), 243–278 (2005)
11. Englert, M., Röglin, H., Vöcking, B.: Worst case and probabilistic analysis of the 2-Opt algorithm for the TSP. Algorithmica **68**(1), 190–264 (2014)
12. Feo, T.A., Resende, M.G.: Greedy randomized adaptive search procedures. J. Global Optim. **6**(2), 109–133 (1995)
13. Festa, P., Resende, M.G.C.: Hybridizations of GRASP with path-relinking. In: Talbi, E.G. (ed.) Hybrid Metaheuristics. Studies in Computational Intelligence, vol. 434, pp. 135–155. Springer, Heidelberg (2013). https://doi.org/10.1007/978-3-642-30671-6_5
14. Fister, I., Yang, X.S., Fister, D., Fister, I.: Cuckoo search: a brief literature review. In: Yang, X.S. (ed.) Cuckoo Search and Firefly Algorithm: Theory and Applications, vol. 516, pp. 49–62. Springer, Cham (2014). https://doi.org/10.1007/978-3-319-02141-6_3
15. Glover, F.: Tabu search-part I. ORSA J. Comput. **1**(3), 190–206 (1989)
16. Glover, F.: Tabu search-part II. ORSA J. Comput. **2**(1), 4–32 (1990)
17. Hansen, P., Mladenović, N.: Variable neighborhood search: principles and applications. Eur. J. Oper. Res. **130**(3), 449–467 (2001)
18. Hart, J., Shogan, A.: Semi-greedy heuristics: an empirical study. Oper. Res. Lett. **6**, 107–114 (1987)

19. Jovanovic, R., Bousselham, A., Voß, S.: Partitioning of supply/demand graphs with capacity limitations: an ant colony approach. J. Comb. Optim. **35**(1), 224–249 (2018)

20. Jovanovic, R., Tuba, M.: An ant colony optimization algorithm with improved pheromone correction strategy for the minimum weight vertex cover problem. Appl. Soft Comput. **11**(8), 5360–5366 (2011)

21. Jovanovic, R., Tuba, M., Voß, S.: An ant colony optimization algorithm for partitioning graphs with supply and demand. Appl. Soft Comp. **41**, 317–330 (2016)

22. Karaboga, D., Gorkemli, B., Ozturk, C., Karaboga, N.: A comprehensive survey: artificial bee colony (ABC) algorithm and applications. Artif. Intell. Rev. **42**(1), 21–57 (2014)

23. van Laarhoven, P.J.M., Aarts, E.H.L.: Simulated annealing. In: van Laarhoven, P.J.M., Aarts, E.H.L., et al. (eds.) Simulated Annealing: Theory and Applications, vol. 37, pp. 7–15. Springer, Dordrecht (1987). https://doi.org/10.1007/978-94-015-7744-1_2

24. Lin, S.: Computer solutions of the traveling salesman problem. Bell Syst. Tech. J. **44**(10), 2245–2269 (1965)

25. Marinakis, Y., Marinaki, M., Dounias, G.: Honey bees mating optimization algorithm for the Euclidean traveling salesman problem. Inf. Sci. **181**(20), 4684–4698 (2011)

26. Marinakis, Y., Migdalas, A., Pardalos, P.M.: Expanding neighborhood GRASP for the traveling salesman problem. Comput. Optim. Appl. **32**(3), 231–257 (2005)

27. Mitchell, M.: An Introduction to Genetic Algorithms. MIT Press, Cambridge (1998)

28. Neri, F., Tirronen, V.: Recent advances in differential evolution: a survey and experimental analysis. Artif. Intell. Rev. **33**(1–2), 61–106 (2010)

29. Reinelt, G.: TSPLIB—a traveling salesman problem library. ORSA J. Comput. **3**(4), 376–384 (1991)

30. Sondergeld, L., Voß, S.: Cooperative intelligent search using adaptive memory techniques. In: Voß, S., Martello, S., Osman, I., Roucairol, C. (eds.) Meta-Heuristics: Advances and Trends in Local Search Paradigms for Optimization, pp. 297–312. Springer, Boston (1999). https://doi.org/10.1007/978-1-4615-5775-3_21

31. Sörensen, K.: Metaheuristics - the metaphor exposed. Int. Trans. Oper. Res. **22**(1), 3–18 (2015). https://doi.org/10.1111/itor.12001

32. Stützle, T., Hoos, H.: Max-min ant system and local search for the traveling salesman problem, pp. 309–314. IEEE (1997)

33. Taillard, E., Voß, S.: POPMUSIC - a partial optimization metaheuristic under special intensification conditions. In: Ribeiro, C., Hansen, P. (eds.) Essays and Surveys in Metaheuristics. Operations Research/Computer Science Interfaces Series, vol. 15, pp. 613–629. Kluwer, Boston (2002). https://doi.org/10.1007/978-1-4615-1507-4_27

34. Tsai, C.F., Tsai, C.W., Tseng, C.C.: A new hybrid heuristic approach for solving large traveling salesman problem. Inf. Sci. **166**(1), 67–81 (2004)

35. Voß, S., Gutenschwager, K.: A chunking based genetic algorithm for the Steiner tree problem in graphs. In: Pardalos, P., Du, D.Z. (eds.) Network Design: Connectivity and Facilities Location. DIMACS Series in Discrete Mathematics and Theoretical Computer Science, vol. 40, pp. 335–355. AMS, Princeton (1998)

36. Woodruff, D.: Proposals for chunking and tabu search. Eur. J. Oper. Res. **106**, 585–598 (1998)

37. Zhu, M., Chen, J.: Computational comparison of GRASP and DCTSP methods for the Traveling Salesman Problem, pp. 1044–1048 (2017)

A Hybrid GRASP/VND Heuristic for the Design of Highly Reliable Networks

Mathias Bourel, Eduardo Canale, Franco Robledo, Pablo Romero,
and Luis Stábile[✉]

Facultad de Ingeniería, Universidad de la República, Montevideo, Uruguay
{mbourel,canale,frobledo,promero,lstabile}@fing.edu.uy

Abstract. There is a strong interplay between network reliability and connectivity theory. In fact, previous studies show that the graphs with maximum reliability, called uniformly most-reliable graphs, must have the highest connectivity. In this paper, we revisit the underlying theory in order to build uniformly most-reliable cubic graphs. The computational complexity of the problem promotes the development of heuristics. The contributions of this paper are three-fold. In a first stage, we propose an ideal Variable Neighborhood Descent (VND) which returns the graph with maximum reliability. This VND works in exponential time. In a second stage, we propose a hybrid GRASP/VND approach that trades quality for computational effort. A construction phase enriched with a Restricted Candidate List (RCL) offers diversification. Our local search phase includes a factor-2 algorithm for an Integer Linear Programming (ILP) model. As a product of our research, we recovered previous optimal graphs from the related literature in the field. Additionally, we offer new candidates of uniformly most-reliable graphs with maximum connectivity and maximum number of spanning trees.

Keywords: Network optimization · Maximum reliability · Heuristics
GRASP · VND · ILP

1 Motivation

In network reliability analysis, the goal is to find the probability of correct operation of a system [2,6]. The context of the original problem determines our notion of correct operation. For instance, delay sensitive applications such as videoconference require a hop-constrained network, where the terminals should be connected by short paths [5]. Wireless systems deal with a hostile environment with mobility (fading, handover and coverage, among other challenges). The goal is to achieve a Grade of Service (GoS) during the busy hour, and node-reliability analysis is more suitable for this context [12]. The interaction between peers in a cooperative environment suggests potential links, and a link-reliability analysis is adequate for this context. Peer-to-peer systems suffer from starvation when the missing-piece syndrome affect all the system [9]. Clearly, the swarm (or population) should be connected, and the all-terminal reliability model is a suitable tool in order to understand this phenomena.

Several researchers from different fields of knowledge (mathematics, computer science, engineering), shaped the body of network reliability analysis, given the application and importance of the underlying models. A fundamental problem is to find the connectedness probability of a random graph, subject to link failures, called the *all-terminal reliability*. The scientific literature around this problem is vast; however, this problem is not fully understood yet. The corresponding practical problem is to connect p sites using q links in the *best* way, this is, to find the graph whose all-terminal reliability is maximum among all (p, q)-graphs. Such graphs are called *uniformly most-reliable graphs*.

The main contributions of this paper are the following:

1. An exact VND that returns uniformly most-reliable graphs is presented.
2. A hybrid GRASP/VND heuristic is introduced in order to find graphs with high reliability. It trades quality for computational feasibility.
3. An Integer Lineal Programming (ILP) formulation called Regularity Problem is proposed. The goal is to find a regular graph starting from a non-regular one moving as minimum number of links as possible.
4. A factor 2 for the Regularity Problem is introduced.
5. Novel networks that show high reliability and connectivity are found, as a result of our hybrid heuristic.

The document is organized in the following manner. Section 2 formally states the problem and breakthroughs in the field of uniformly most-reliable graphs. Section 3 presents an exact VND that runs in exponential time, and a hybrid GRASP/VND heuristic that trades quality for computational feasibility. As a product, we offer novel cubic networks with high reliability in Sect. 4. Concluding remarks and open problems are discussed in Sect. 5.

2 Uniformly Most-Reliable Graphs

2.1 Definition

In the following, we work with undirected graphs without loops, and a graph with p nodes and q links is a (p, q)-graph.

Definition 1. *Consider a graph G with perfect nodes but independent link failures with identical probability $\rho \in (0, 1)$. The all-terminal reliability, $R_G(\rho)$, is the probability that the resulting subgraph remains connected.*

The *unreliability* $U_G(\rho) = 1 - R_G(\rho)$ can be expressed using sum-rule:

$$U_G(\rho) = \sum_{k=0}^{q} m_k(G)\rho^k(1-\rho)^{q-k}, \qquad (1)$$

being $m_k(G)$ the number of spanning disconnected subgraphs of G with exactly $q - k$ links. Therefore, $R_G(\rho)$ is a polynomial in $\rho \in (0, 1)$, and its determination is reduced to counting the numbers $\{m_k\}_{k=0,...,q}$.

Definition 2. *A (p,q)-graph H is* uniformly most-reliable *if $R_H(\rho) \geq R_G(\rho)$ for all (p,q)-graph G and all $\rho \in (0,1)$.*

Alternatively, H is uniformly most-reliable if its unreliability $U_H(\rho)$ is dominated (i.e., upper-bounded) by all functions $U_G(\rho)$ for all (p,q)-graph G.

2.2   Breakthroughs

In this section we present fundamental results that are the cornerstone in the theory of uniformly most-reliable graphs. The following section briefly describes the main findings that complement the fundamental results.

In 1977, Arnie Rosenthal formally proved that the K-terminal reliability evaluation belongs to the class of $\mathcal{NP}$-Hard computational problems [19]. The key concept of the proof is the reducibility introduced in 1972 by Richard Karp, which represents a foundational work in computational complexity [13]. As corollary, finding uniformly most-reliable graphs is a hard problem as well.

Observe that if $m_k(H) \leq m_k(G)$ for all $k \in \{0, \ldots, q\}$ and (p,q)-graph G, then H is uniformly most-reliable. This is a simple but elegant interplay between network reliability analysis and connectivity theory. Curiously enough, the converse is still an open problem:

Conjecture 1 (Boesch et al.). If G is uniformly most-reliable (p,q)-graph, then $m_k(G) \leq m_k(H)$ for all (p,q)-graph H.

If $\lambda(H)$ denotes the connectivity of H and $\tau(H)$ its number of spanning trees, the following necessary criterion holds [1]:

Corollary 1. *A uniformly most-reliable graph H must have the maximum tree-number $\tau(H)$, maximum connectivity $\lambda(H)$, and the minimum number $m_\lambda(H)$.*

Corollary 1 wakes up interest in two special sub-problems: the maximum connectivity and maximum tree-number of a graph. In the second book ever written in graph theory, Claude Berge challenges the readers to find the graph with maximum connectivity among all graphs with a fixed number of nodes and links. Frank Harary provided not only a full answer, but also found connected graphs with minimum and maximum diameter [10]. The idea behind his construction is simple: by handshaking, the average degree of a (p,q)-graph is $\frac{2q}{p}$. Therefore, $\lambda \leq \lfloor \frac{2q}{p} \rfloor$. Harary graphs achieve this upper-bound, which represents the maximum connectivity of a graph.

Gustav Kirchhoff solved linear time-invariant resistive circuits, and as corollary he introduced the Matrix-Tree theorem, where he counts the number of spanning trees of a connected graph (i.e., the tree-number) using the determinant of a matrix [3,15]. This breakthrough in electrical systems launched the theory of trees, which provides the building blocks in communication design. However, the corresponding extremal problem is not well understood: find the graph with a fixed number of nodes and links that maximizes the tree-number.

For convenience we say that a (p, q)-graph, H, is t-optimal if $\tau(H) \geq \tau(G)$ for every (p, q) graph G. Briefly, Corollary 1 claims that uniformly most-reliable graphs must be t-optimal and max-λ min-m_λ, where λ denotes the edge connectivity.

Another breakthrough from the related literature is a reliability improving graph transformation called *swing surgery*, independently discovered by Kelmans [14] and Satyanarayana et al. [20]. Specifically, if we are given a (p, q)-graph $G = (V, E)$, two nodes $x, y \in V$ with respective neighboring nodes S_x and S_y with $S_x \setminus \{y\} \subset S_y$, $B \subseteq S_y - \{x\}$ such that $B \cap S_x = \emptyset$, and G' the graph obtained by G by removing the links $\{(y, z), z \in B\}$ and adding the links $\{(x, z), z \in B\}$, then $R_{G'}(\rho) \geq R_G(\rho)$ for all $\rho \in (0, 1)$.

2.3 Findings

Redundancy is of paramount importance in communication networks. For that reason, in the following we are specifically focused on uniformly most-reliable cubic graphs (i.e., 3-regular connected graphs). By handshaking, this is the case of $(2r, 3r)$-graphs for some $r \geq 2$. Recall that Möbius graph M_n is precisely the elementary cycle C_{2n} together with all the diameters (opposite nodes are also linked). So far, the findings of uniformly most-reliable cubic graphs can be summarized in the following list:

- $K_4 = M_2$; complete graph with 4 nodes; case $r = 2$; see [4].
- $K_{(3,3)} = M_3$; complete bipartite graph; case $r = 3$; see [23].
- $W_4 = M_4$; Wagner graph ($r = 4$); see [18].
- P_5; Petersen graph ($r = 5$); see [16].
- Y_6; Yutsis graph ($r = 6$); see [22].

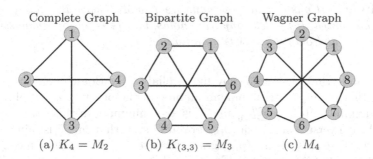

Fig. 1. Möbius graphs M_2, M_3 and M_4

The reader is invited to consult the corresponding references for a mathematical proof that these graphs are uniformly most-reliable. They are sketched in Figs. 1 and 2. By computational limits, currently it is not possible to find uniformly

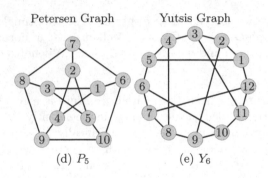

Petersen Graph   Yutsis Graph

(d) P_5 (e) Y_6

Fig. 2. Petersen P_5 and Yutsis graph Y_6.

most-reliable cubic graphs for $r \geq 7$. The goal of this paper is to build highly reliable cubic graphs for $r \in \{7, \ldots, 10\}$, finding a trade-off between quality and computational effort.

2.4 Equivalent Combinatorial Problem

It is worth to remark that the problem of finding uniformly most-reliable graphs is a simultaneous minimization of an uncountable family of numbers $\{U_G(\rho)\}_{\rho \in (0,1)}$. However, if Boesch Conjecture holds, we observe that the problem can be translated to a (single-objective) combinatorial optimization problem. Specifically, let us denote $m(G)$ to the number of disconnected spanning subgraphs for G. By the definition of link disconnecting sets, we get that:

$$m(G) = \sum_{k=0}^{q} m_k(G).$$

Proposition 1. *Consider natural numbers p and q such that there exists a unique (p,q)-graph. If Boesch conjecture holds, then G is uniformly most-reliable (p,q) graph if and only if $m(G)$ is minimum.*

Proof. Assume that G is uniformly most-reliable. By Boesch conjecture, every disconnecting set $m_k(G)$ is minimum among all the other (p,q)-graphs. Therefore, the number $m(G) = \sum_{k=0}^{q} m_k(G)$ is also minimum in this set.

For the converse, consider a (p,q)-graph G such that $m(G)$ is minimum. By hypothesis, there exists some (p,q)-graph, denoted by H. By Boesch conjecture, $m_k(H) \leq m_k(G)$. Since $m(G) = \sum_{k=0}^{q} m_k(G)$ is minimum, the only possibility is that $m_k(G) = m_k(H)$ for all k. Therefore, G and H share the same unreliability polynomial. By uniqueness, we must have $G = H$, and the statement is proved.

In short, Proposition 1 tells us that if Boesch conjecture holds, then finding uniformly most-reliable graphs is equivalent to the minimization of disconnected spanning subgraphs $m(G)$. This result reinforces the evidence that the optimum graphs under connectivity (i.e., purely deterministic) and reliability optics (probabilistic) share common properties.

In this work we are focused on the minimization of disconnecting spanning subgraphs $m(G)$. In this paper, we offer highly reliable cubic graphs, which share strong connectivity properties as well, supported by Proposition 1.

It is well-known that finding the coefficients $\{m_k(G)\}_{k=0,\ldots,q}$ belongs to the hierarchy of #$\mathcal{P}$-Complete counting problems [21]. Furthermore, the number $m(G) = \sum_{k=0}^{q} m_k(G) = T(1,2)$, is precisely Tutte polynomial evaluated at the point $(1,2)$, which is a #$\mathcal{P}$-Hard counting problem, even for bipartite planar graphs [11]. Here, we propose a pointwise statistical estimation of this number. Monte Carlo is a noteworthy computational tool for simulation. From a macroscopic point of view, the idea is to faithfully simulate a complex system (or a part of it), and consider N independent experiments of that simulation, in order to determine the performance of the system (or subsystem) and assist decisions on it [8].

We will use Crude Monte Carlo (CMC) in order to provide an unbiased statistical estimation for $m(G)$. First of all, observe that $m(G)$ is strictly related with the unreliability evaluation at $\rho = \frac{1}{2}$:

$$U_G(\frac{1}{2}) = \sum_{k=0}^{q} m_k(G)(\frac{1}{2})^k (\frac{1}{2})^{q-k} = \frac{m(G)}{2^q}. \tag{2}$$

Equation 2 shows that, alternatively, we must minimize $U_G(\frac{1}{2})$, or the probability that the resulting subgraph is disconnected under identical independent link failures with probability $\rho = \frac{1}{2}$. For any given graph G, let us consider a sample of random graphs $G_1, \ldots, G_N$ picked independently with link failures $\rho = \frac{1}{2}$, and independent Bernoulli variables $X_1, \ldots, X_N$ such that $X_i = 1$ if and only if G_i is disconnected. By strong law of large numbers, the mean sample $\overline{X_N}$ converges almost surely to $u = U_G(\frac{1}{2})$. Therefore, in order to decide whether $m(G_1) < m(G_2)$ or not, we use the criterion $\overline{X_N}^1 < \overline{X_N}^2$ for N large enough, being $\overline{X_N}^i$ the mean sample for the graph G_i. This criterion avoids the full determination of the coefficients $m_k(G)$, and it will be useful for the design of a GRASP/VND heuristic to build highly reliable graphs.

3   Metaheuristics

In this section we develop an ideal VND metaheuristic that returns a uniformly most-reliable graph. Since it has exponential time, we must trade accuracy for computational effort. As a consequence, a full GRASP/VND heuristic is introduced.

3.1   VND

Variable Neighborhood Descent (VND) explores several neighborhood structures in a deterministic order. Its success is based on the simple fact that different neighborhood structures do not usually have the same local minima. Thus, the local optima trap problem is addressed by a deterministic change of neighborhoods [7].

Recall that a simultaneous minimization of the coefficients $\{m_k(G)\}_{k=0,\ldots,q}$ is a sufficient condition for G to be uniformly most-reliable. Therefore, if there is one local search dedicated to each coefficient, the output must be uniformly most-reliable. Trivial neighborhood structures where all (p,q)-graphs are neighbors of some fixed graph work. However, the cardinality of the search-space of (p,q)-graphs is $\binom{p(p-1)/2}{q}$. Therefore, an exhaustive search among the trivial neighborhood structures of all (p,q)-graphs is computationally prohibitive.

3.2 GRASP/VND Heuristic

GRASP is an iterative multi-start process which operates in two phases [17]. In the Construction Phase a feasible solution is built whose neighborhood is then explored in the Local Search Phase [17]. The second phase is usually enriched by means of different variable neighborhood structures, for instance, VND.

We adapt the previous ideal VND in order to obtain a feasible computational solution in a multi-start fashion with diversification in a previous construction phase. Algorithm $HighlyReliable$ receives a maximum number of iterations $iter$, a natural number $r \geq 2$, and returns a highly reliable cubic $(2r, 3r)$-graph.

Algorithm 1. $G = HighlyReliable(r, iter)$

1: $G \leftarrow M_r$
2: **for** $i = 1$ to $iter$ **do**
3:     $G_{input} \leftarrow GreedyRandomized(r, \alpha)$
4:     $G(i) \leftarrow VND(G_{input})$
5:     **if** $m(G(i)) \leq m(G)$ **for all** k **then**
6:         $G \leftarrow G(i)$
7:     **end if**
8: **end for**
9: **return** G

In Line 1, the graph is initialized in Möbius graph M_r, which is known to be optimal for the cases where $r \in \{2, 3, 4\}$. In a **for**-loop with $iter$ iterations (Lines 2–8), we iteratively call in sequence the Construction Phase (Line 3) and VND (Line 4). If the number of disconnecting spanning subgraphs $m(G(i))$ is dominated by $m(G)$, the current graph G is replaced by $G(i)$ (Lines 5–6). It is worth to remark that the test $m(G(i)) \leq m(G)$ considers the criterion detailed in Subsect. 2.4. The best graph among all the iterations is returned as the output (Line 9).

In the following, we provide details of the Construction Phase ($GreedyRandomized(r)$ from Line 3) and Local Search Phase (VND function, from Line 4). The result is not necessarily a uniformly most-reliable network, but a highly reliable cubic network, which is useful for practical purposes.

Algorithm 2. $G = Construct(r)$

1: $U \leftarrow RandomNumbers(r(r-1)/2)$
2: $G \leftarrow RandomTree(U)$
3: $\delta \leftarrow \min_{v \in G}\{deg(v)\}$
4: $\Delta \leftarrow \max_{v \in G}\{deg(v)\}$
5: $RCL \leftarrow \{(v_i, v_j) : deg(v_i)deg(v_j) \leq \delta^2 + \alpha(\Delta^2 - \delta^2)\}$
6: **for** $i = 1$ to $r + 1$ **do**
7:     $e_i \leftarrow Random(RCL)$
8:     $G \leftarrow G \cup \{e_i\}$
9:     $RCL \leftarrow Update(RCL, e_i)$
10: **end for**
11: **return** G

Construction Phase. The main idea is to start with a random tree with $2r - 1$ links and insert adequately $r + 1$ links meeting a final size of $3r$ links. In Line 1, for every pair of potential links we pick independent numbers in $(0, 1)$ uniformly chosen at random. In this way, we get random costs c_{ij} for every pair of nodes v_i and v_j. A random tree is found in Line 2. Specifically, function *RandomTree* applies Kruskal algorithm with the costs c_{ij}. The minimum and maximum degree of the resulting graph are found in Lines 3 and 4 respectively. The addition of the remaining $r + 1$ links takes place in the block of Lines 5–10. A Restricted Candidate List, RCL, selects a percentage of α links $e_{(i,j)}$ with the lowest product degrees $deg(v_i) \times deg(v_j)$; see Line 5. In the **for**-loop of Lines 6–10, links are iteratively picked from the RCL (Line 7) and added to the graph G (Line 8). Observe that the RCL should be updated, since the degrees are modified in each iteration. This operation takes place in function $Update(RCL, e_i)$ (Line 9). Clearly, G is not necessarily regular, but the effect of the RCL provides diversity in the solutions. Naturally, it trades greediness for randomization with the parameter α, and tends to return almost-regular graphs.

Local Search Phase. In the Local Search Phase, a VND is considered with the following movements:

1. *Surgery*: applies the graph transformation called Swing Surgery (see Subsect. 2.2).
2. *Regular*: returns a regular graph after adequate link addition/deletions.
3. *Crossing*: tests whether the tree-number is increased after all feasible graph-crossings.

A graph G is *healthy* if there is no feasible surgery that improves the reliability uniformly in $(0,1)$. In other words, it is locally optimum with respect to local movements under Swing Surgery. Analogously, we say that G is *strong* if G

has the largest tree-number with respect to all feasible crossings (i.e., it is a locally optimum solution with respect to local movements of *Crossing*). Figure 3 presents the full VND. The reader can observe that the output G is regular, healthy and strong.

In the following, we explain the three movements. Let us start with *Surgery* and *Crossing* (as we will see, *Regular* movement includes an ILP formulation, a new result on approximation algorithms and an exact polynomial time algorithm to solve it). *Surgery* just applies a reliability-improving graph transformation called Swing Surgery from Subsect. 2.2 whenever possible. Finally, *Crossing* tries to find an edge-crossing with largest tree-number. Specifically, if the links $e_1 = (x, y)$ and $e_2 = (z, t)$ belong to G, but $(x, z), (y, t)$ do not belong to G, *Crossing* counts the tree-number of the new graph $G' = (G - \{(x, y), (z, t)\}) \cup \{(x, z), (y, t)\}$. The tree-number is efficiently found by Kirchhoff theorem, as any cofactor of the Laplacian matrix [3].

The main idea of *Regular* is to return a regular graph, starting from a non-regular one. *Regular* movement is a solution to an Integer Linear Programming formulation. Consider the input graph $G = (V, E)$, the resulting regular graph $G' = (V, E')$ and the following binary variables:

- $e_{(i,j)} = 1$ iff $(i, j) \in E$ (adjacency matrix for G);
- $a_{(i,j)} = 1$ iff $(i, j) \in E' - E$ (links added to G');
- $r_{(i,j)} = 1$ iff $(i, j) \in E - E'$ (links removed from G).

Our goal is to minimize the number of addition/deletions in order to return a 3-regular graph G'. The Regularity Problem can be formalized by the following ILP:

$$\min \sum_{i<j} a_{(i,j)} \tag{3}$$

$$s.t. \tag{4}$$

$$\sum_{i<j} a_{(i,j)} = \sum_{i<j} r_{(i,j)} \tag{5}$$

$$\sum_{i<j} a_{(i,j)} + e_{(i,j)} - r_{i,j} = 3 \, \forall i \in \{1, \ldots, 2r\} \tag{6}$$

$$a_{(i,j)}; r_{(i,j)} \in \{0, 1\} \, \forall i < j. \tag{7}$$

Constraint 5 states that the number of added/removed links must be identical, so, $|E| = |E'|$. Constraint 6 state that the resulting graph G' must be 3-regular. Finally, Constraint 7 determine the binary domain for the decision variables $a_{(i,j)}$ and $r_{(i,j)}$.

In the following, *Regular* movement is specified.

Algorithm 3. $G = Regular(G)$

1: $\Delta \leftarrow \max_{v \in V(G_{in})}\{deg(v)\}$
2: $\delta \leftarrow \min_{v \in V(G_{in})}\{deg(v)\}$
3: **while** $\delta(G) < \Delta(G)$ **do**
4:　　$x \leftarrow \arg\max_{u \in V}\{deg(u)\}$
5:　　$y \leftarrow \arg\min_{u \in V}\{deg(v')\}$
6:　　$z \leftarrow Random(N(x) - N(y))$
7:　　$G \leftarrow (G - (x,z)) \cup (y,z)$
8: **end while**
9: **return** G

Regular function receives a (p,q)-graph G and returns a regular graph. The key is to move links from nodes with the highest degree to nodes with the lowest degree. In Lines 1–2, the maximum and minimum degrees for the input graph are found. A **while**-loop (Lines 3–8) takes effect whenever $\delta < \Delta$. Since the degree of some low-degree (high-degree) node is increased (resp. decreased) by a unit in all the iterations, the number of iterations is finite, and the algorithm returns a 3-regular graph. In Lines 4 and 5, we pick some node x (y) with the highest (resp. lowest) degree. Since $deg(x) > deg(y)$, there exists some node, z, such that z is adjacent to x but non-adjacent to y (Line 6). In the iteration, the link (x,z) is deleted, while (y,z) is added (Line 7). Observe that, in the resulting graph, the degree of x (y) is decreased (resp. increased), but the degree of the pivotal node z is identical. The output of **while**-loop must be a regular graph, which is returned in Line 9.

Theorem 1. *Regular is an approximation algorithm with factor 2 for the Regularity Problem.*

Proof. In an arbitrary link addition/deletion, we can reduce at most 2 degrees of high-degree nodes and increase 2 degrees in the set of low-degree nodes. *Regular* performs methodically an addition/reduction of a single-degree in each movement. Therefore, the number of movements during the execution of *Regular* is, at most, twice the optimal solution. □

4   Results

By construction, it is clear that our GRASP/VND heuristic returns M_r for the known cases $r \in \{2,3,4\}$ (see Line 1 of *Construction*). In order to test the effectiveness of our GRASP/VND heuristic, we look for $r \in \{5,\ldots,10\}$. We know beforehand that the uniformly most reliable graph for $r = 5$ is Petersen graph [16]; when $r = 6$ Yutsis graph Y_6.

Figure 4 sketches the six resulting graphs for $r \in \{5,6,7,8,9,10\}$ using $iter = 10^5$ iterations, $\alpha = 0.5$ and sample size $N = 10^4$ for Crude Monte Carlo for the pointwise reliability estimation of $U_G(\frac{1}{2})$ as detailed in Subsect. 2.4.

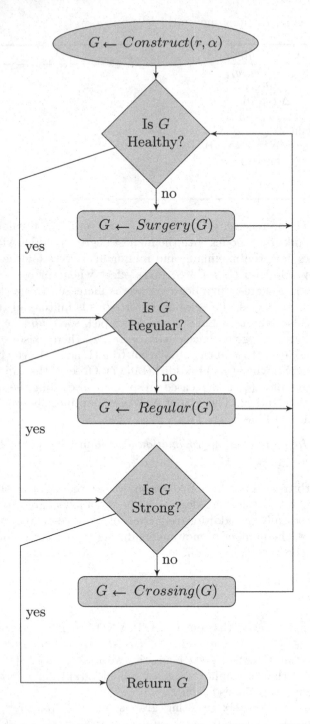

Fig. 3. Flow diagram for the Local Search Phase - VND.

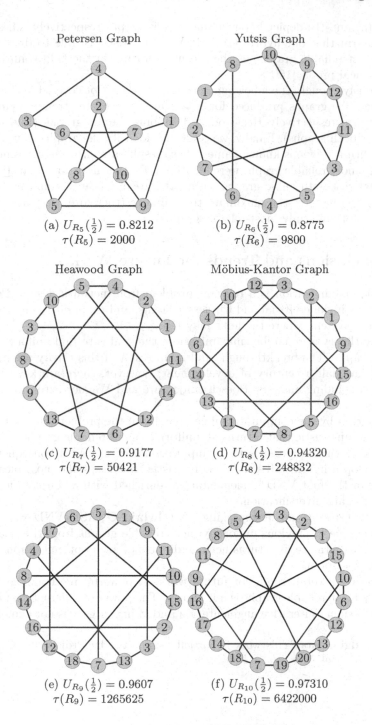

Fig. 4. Results for $r \in \{5, 6, 7, 8, 9, 10\}$

Figure 4(a) and (b) depict Petersen and Yutsis graphs respectively. These base-steps confirm that our hybrid GRASP/VND heuristic is able to discover uniformly most-reliable graphs, as the recent literature in the field confirms with mathematical proofs [16].

Curiously enough, the following cases for $r \geq 7$ are not covered in the related literature. The graphs produced for $r = 7$ and $r = 8$ are strongly symmetric, and they are respectively Heawood and Möbius-Kantor graphs (Fig. 4(c) and (d) depict both graphs). Finally, we could not identify the graphs from Fig. 4(e) and (f) with a previous known name. These results are encouraging, since some uniformly most-reliable graphs were identified. Furthermore, an exhaustive computational test with cubic graphs with girth greater than 3 confirms that the resulting graphs have the maximum tree-number (therefore, they are the only candidates of uniformly most-reliable graphs).

5   Conclusions and Trends for Future Work

Reliability maximization is a relevant problem from network design. Potential applications include virtual and wireless systems, and cooperative environments in a hostile system, where the links may fail.

In the theory of reliability maximization, the goal is to find uniformly most-reliable graphs. The breakthroughs and main result of this theory are here outlined. It has half a century of development; however, there are key questions without concluding answers. Boesch conjecture and Wagner extension are just examples.

Supported by the computational complexity of the problem, we first present a VND metaheuristic that returns all uniformly most-reliable graphs. The main drawback of this proposal is the computational effort. As a consequence, we then develop a hybrid GRASP/VND heuristic that keeps the most meaningful elements of the first VND implementation, enriched with a construction phase in order to gain diversification.

The first results are encouraging. Our hybrid GRASP/VND was able to detect a couple of previous uniformly most-reliable graphs from the related literature. Furthermore, it returns new candidates of such optimal graphs from a reliability viewpoint.

There are several trends for future work. A powerful methodology to find uniformly most-reliable graphs is not known. The power of different graph transformations such as iterative augmentation and swing surgery is not explored yet.

Acknowledgements. This work is partially supported by Project 395 CSIC I+D *Sistemas Binarios Estocásticos Dinámicos.*

References

1. Bauer, D., Boesch, F., Suffel, C., Van Slyke, R.: On the validity of a reduction of reliable network design to a graph extremal problem. IEEE Trans. Circuits Syst. **34**(12), 1579–1581 (1987)
2. Beineke, L.W., Wilson, R.J., Oellermann, O.R.: Topics in Structural Graph Theory. Encyclopedia of Mathematics and its Applications. Cambridge University Press, Cambridge (2012)
3. Biggs, N.: Algebraic Graph Theory. Cambridge Mathematical Library. Cambridge University Press, Cambridge (1993)
4. Boesch, F.T., Li, X., Suffel, C.: On the existence of uniformly optimally reliable networks. Networks **21**(2), 181–194 (1991)
5. Canale, E., Cancela, H., Robledo, F., Romero, P., Sartor, P.: Full complexity analysis of the diameter-constrained reliability. Int. Trans. Oper. Res. **22**(5), 811–821 (2015)
6. Colbourn, C.J.: Reliability issues in telecommunications network planning. In: Sansò, B., Soriano, P. (eds.) Telecommunications Network Planning. CRT, pp. 135–146. Springer, Boston (1999). https://doi.org/10.1007/978-1-4615-5087-7_8
7. Duarte, A., Mladenović, N., Sánchez-Oro, J., Todosijević, R.: Variable neighborhood descent. In: Martí, R., Panos, P., Resende, M. (eds.) Handbook of Heuristics, pp. 1–27. Springer, Heidelberg (2016). https://doi.org/10.1007/978-3-319-07153-4_9-1
8. Fishman, G.S.: Monte Carlo: Concepts, Algorithms and Applications. Springer, New York (1996). https://doi.org/10.1007/978-1-4757-2553-7
9. Hajek, B., Zhu, J.: The missing piece syndrome in peer-to-peer communication. In: 2010 IEEE International Symposium on Information Theory, pp. 1748–1752, June 2010
10. Harary, F.: The maximum connectivity of a graph. Proc. Natl. Acad. Sci. U. S. A. **48**(7), 1142–1146 (1962)
11. Jaeger, F., Vertigan, D.L., Welsh, D.J.A.: On the computational complexity of the Jones and Tutte polynomials. In: Mathematical Proceedings of the Cambridge Philosophical Society, vol. 108, no. 1, pp. 35–53 (1990)
12. Jin, R., et al.: Detecting node failures in mobile wireless networks: a probabilistic approach. IEEE Trans. Mob. Comput. **15**(7), 1647–1660 (2016)
13. Karp, R.M.: Reducibility among combinatorial problems. In: Miller, R.E., Thatcher, J.W. (eds.) Complexity of Computer Computations, pp. 85–103. Plenum Press, New York (1972)
14. Kelmans, A.K.: On graphs with randomly deleted edges. Acta Math. Acad. Sci. Hung. **37**(1), 77–88 (1981)
15. Kirchoff, G.: Über die auflösung der gleichungen, auf welche man bei der untersuchung der linearen verteilung galvanischer ströme geführt wird. Ann. Phys. Chem. **72**, 497–508 (1847)
16. Rela, G., Robledo, F., Romero, P.: Petersen graph is uniformly most-reliable. In: Nicosia, G., Pardalos, P., Giuffrida, G., Umeton, R. (eds.) MOD 2017. LNCS, vol. 10710, pp. 426–435. Springer, Cham (2018). https://doi.org/10.1007/978-3-319-72926-8_35
17. Resende, M.G.C., Ribeiro, C.C.: Optimization by GRASP - Greedy Randomized Adaptive Search Procedures. Springer, New York (2016). https://doi.org/10.1007/978-1-4939-6530-4

18. Romero, P.: Building uniformly most-reliable networks by iterative augmentation. In: 9th International Workshop on Resilient Networks Design and Modeling (RNDM), pp. 1–7 (2017)
19. Rosenthal, A.: Computing the reliability of complex networks. SIAM J. Appl. Math. **32**(2), 384–393 (1977)
20. Satyanarayana, A., Schoppmann, L., Suffel, C.L.: A reliability improving graph transformation with applications to network reliability. Networks **22**(2), 209–216 (1992)
21. Provan, J.S., Ball, M.O.: The complexity of counting cuts and of computing the probability that a graph is connected. SIAM J. Comput. **12**(4), 777–788 (1983)
22. Viera, J.: Búsqueda de grafos cúbicos de máxima confiabilidad. Master's thesis, Facultad de Ingeniería, Universidad de la República, Uruguay (2018)
23. Wang, G.: A proof of Boesch's conjecture. Networks **24**(5), 277–284 (1994)

Game of Patterns and Genetic Algorithms Under a Comparative Study

Ebert Brea[(✉)] [iD]

Facultad de Ingeniería, Escuela de Ingeniería Eléctrica, Dpto. Electrónica,
Computación y Control, Universidad Central de Venezuela, Caracas, Venezuela
ebert.brea@ucv.ve, ebertbrea@gmail.com,
https://www.researchgate.net/profile/Ebert_Brea

Abstract. In this article we present a comparison of the performance
between a metaheuristic optimization method, Game of Patterns (GofP),
so-called by the author, and the well-known genetic algorithms (GAs),
through two implementations, namely: the GA of Scilab (SGA); and
the GA of the R Project for Statistical Computing (RGA). For this
purpose, we have selected a set of multimodal objective functions in
the n-dimensional Euclidean space $\mathbb{R}^n$ with a unique global minimum.
For comparing both metaheuristic optimization approaches, a performance indicator of quality, denoted $Q(p, n, s)$, was defined, which allows
us to measure the quality of the obtained global optimal solution for
each pth problem, in the n-dimensional space, when it is solved by each
metaheuristic optimization method $s \in \{\texttt{GofP}, \texttt{SGA}, \texttt{RGA}\}$. The indicator
$Q(p, n, s)$ then depends on: the number of evaluations of the pth opti-
mization problem in the Euclidean space $\mathbb{R}^n$, which has required the s
metaheuristic optimization method for identifying the global minimum;
and the distance between the location of its respective unique global minimum and the location of the minimum that has been identified by the s
metaheuristic optimization method. The paper also offers a brief explanation of the GofP method, which has been developed for solving unconstrained mixed integer problems in the $n \times m$-dimensional Euclidean
space $\mathbb{R}^n \times \mathbb{Z}^m$.

Keywords: Game of Patterns · Genetic algorithms
Comparison of metaheuristic optimization methods

1 Introduction

Consider the following unconstrained nonlinear problem:

$$\underset{x \in \mathbb{R}^n}{\text{minimize}}\, f(x), \tag{1}$$

where $f(x) : \mathbb{R}^n \to \mathbb{R}$ is a nonlinear objective function with just one global
minimum. Furthermore, we have also assumed that an explicit mathematical
expression of $f(x)$ is not necessarily available.

© Springer Nature Switzerland AG 2019
M. J. Blesa Aguilera et al. (Eds.): HM 2019, LNCS 11299, pp. 93–107, 2019.
https://doi.org/10.1007/978-3-030-05983-5_7

A metaheuristics method, so called Game of Patterns (GofP), has been recently developed by the author for globally solving mixed integer nonlinear problems [4,5], which is been here applied to some numerical examples in the n-dimensional Euclidean space $\mathbb{R}^n$ for comparing it with the well known genetic algorithms (GAs).

The author has obtained a very good performance when the GofP method is applied to unconstrained mixed integer nonlinear optimization problems, for identifying at least a global solution. Moreover, the GofP method has shown to have an excellent performance when it is also applied to constrained mixed integer nonlinear optimization problems. In this last case, the penalty function approach has been applied for solving this case. The GofP method is based on a zero-sum game framework, which was originally developed for globally solving mixed integer nonlinear problems [4,5]. The GofP method is conformed by a set of player. Each player carries out one iteration, at each round, of a search algorithm called Mixed Integer Randomized Pattern Search Algorithm (MIRPS), developed by Brea [3].

The main purpose of this paper is to show the results of statistically comparison between the GofP method and the implementations of the genetic algorithms under: Scilab and the language and environment for statistical computing R for solving minimization of an objective function under bounded constraints [2,11,12].

The remainder of this article is organized as follows. A brief explanation of the GofP method is discussed in Sect. 2. In Sect. 3 we also offer a short explanation on the GAs. In Sect. 4, we describe the performance measure defined in this research for comparing both metaheuristic optimization approaches. In Sect. 5 we show a set of nonlinear numerical problems and statistical summaries of performance indicators for statistically comparing the metaheuristic approaches. Finally, conclusions and future research are discussed in Sect. 6.

2   The Game of Patterns

Let $\mathcal{P}$ be the following unconstrained mixed integer nonlinear optimization problem,

$$\underset{(x,y)\in\mathbb{R}^n\times\mathbb{Z}^m}{\text{minimize}} \; f(x,y), \tag{2}$$

where $f(x,y) : \mathbb{R}^n \times \mathbb{Z}^m \to \mathbb{R}$ is a mixed integer nonlinear objective function. Furthermore, note that an explicit mathematical expression of $f(x,y)$ is not necessarily available, which could be evaluated at each point $(x^t, y^t)^t$ by solving a system of equations or by simulation models.

The main idea of the Game of Patterns (GofP) method for solving the problem $\mathcal{P}$ is based on a zero-sum game framework, wherein there are η players, and each one of them is a randomized pattern search algorithm. In our case, each ℓth player performs just one iteration of a novel random search algorithm for solving mixed integer nonlinear problems, which is called the Mixed Integer Randomized Pattern Search Algorithm (MIRPSA) [3]. The mixed integer randomized pattern search of the MIRPSA is modified by two main operations of the MIRPSA for finding at least a local minimum of the problem $\mathcal{P}$. These main operations are: moving operation and shrinking operation. Each operation is carried out by the MIRPSA when a set of conditions are present [3].

Therefore, the GofP method arranges the game among the set of mixed integer randomized pattern searches $\mathcal{H}_\ell^{[k]}$ at each kth game round, where each ℓth player accounts with an initial balance $b_\ell^{[0]}$ at the beginning of the game, this is at the beginning of the 1st round. Namely, let $\Gamma = \{\mathcal{H}_\ell; \mathcal{S}_\ell; \mathcal{Q}_\ell\}_{\ell=1}^{\eta}$ be a framework of a zero-sum game defined by a set of η players $\{\mathcal{H}_\ell\}_{\ell=1}^{\eta}$, wherein each ℓth player has a set of $\mathcal{S}_\ell$ strategies, and a pay-off function for each ℓth player, which is given by $\mathcal{Q}_\ell : \times_{\ell=1}^{\eta} \mathcal{S}_\ell \rightarrow \mathbb{N}$, [7].

The strategy of each ℓth player at each kth round is given by a random number of trial points, which is given by a uniform distribution between $n + m$ and $2(n + m)$, within its mixed integer randomized pattern search $\mathcal{H}_\ell^{[k]}$. The winner of each kth round is the player that had identified the best value of the objective function. In this case, each loser must pay the winner an among equal to its number of trial points. We must point out that, according to the rule of the GofP, any ℓth active player $\mathcal{H}_\ell^{[k]}$ will become a disqualified player $\overline{\mathcal{H}}_\ell^{[k]}$ at any $(k + 1)$th round, if his account balance $b_\ell^{[k+1]}$, after the kth round, becomes less than the minimum level of bet $M_\ell = n + m$, then it will cause a disqualification of the player, and therefore this player will not come back to the game.

This process is recurrently repeated until that just leaves one qualified player, who will be considered the winner of the game. At this stage, the winner of the game restarts the MIRPSA using its last location as starting point for accurately identifying the minimum of the problem.

A study of some convergence properties are presented by Brea [5], when the GofP method is applied to unconstrained mixed integer nonlinear problems. However, the GofP method can be also applied to constrained problems, using the penalty function approach, what this viewpoint also allows us to identify at least a global solution of the constrained problem under minimization.

Despite the GofP method was developed for solving mixed integer nonlinear problems [5], the GofP method can be applied to nonlinear problems only defined either in the real field $\mathbb{R}^n$ or in the integer field $\mathbb{Z}^m$. We consequently need pointing out that for this study, we have applied the GofP method to a set of test problems uniquely defined in the Euclidean real field $\mathbb{R}^n$.

3   The Genetic Algorithms

Genetic algorithms (GAs) are considered metaheuristic and bioinspirited random search algorithms, originally developed by Holland [6], and they have been applied for globally solving nonlinear optimization problems.

The main idea of GAs are based on the adaptive processes associated with natural genetics, wherein with an initial set of random potential solutions called the population (generation) are evaluated using the objective function of the optimization problem being solved. In each generation, among the best individuals or solutions are randomly selected for modifying its genome by recombination and possibly random mutations, who will yield a new generation. The individuals of the new generation are evaluated using the objective function for carrying out a next iteration. The algorithm commonly finishes by number of generations.

There exists a vast amount of literature on the GAs, for instance, [1,8,14]. We have used two implementations of the GAs for solving the minimization of an objective function subject to bounded constraints, which both are part of the toolboxes of the free and open source softwares: Scilab [11] and R [10]. The Scilab GA (SGA) was included in the 5th Scilab version thanks to the contributions of Collette [2]. Whilst, the GA by R (RGA) was implemented by Scrucca [12].

Both implementations, the SGA and the RGA, were developed for globally solving bounded nonlinear problems. Namely, for solving nonlinear problems such as:

$$\operatorname*{minimize}_{x \in \mathbb{R}^n} f(x), \tag{3a}$$

$$\text{subject to} : l^{(k)} \le x^{(k)} \le u^{(k)}, \quad \forall k \in \{1, \ldots, n\}, \tag{3b}$$

where $f(x) : \mathbb{R}^n \to \mathbb{R}$ is a nonlinear objective function, and $l^{(k)}$ and $u^{(k)}$ are respectively the kth lower and kth upper bounds for each kth component of the vector $x = (x^{(1)}, \ldots, x^{(n)})^t$.

4   Performance Measure of Comparison

Our target in this research is to measure the performance of the GofP method versus both the SGA and the RGA for comparing them, using a representative number of samples or running replications, each with different random number sequences for each metaheuristic method.

For comparing these sets of observations, which were respectively yielded by the performance measure of each metaheuristic method, we have then used a measure of quality, for each pth test problem, in the n-dimensional real Euclidean space, each s metaheuristic method, given by $s \in \{\texttt{GofP}, \texttt{SGA}, \texttt{RGA}\}$, and each ith replication or sample of the respective metaheuristic method. Namely, each ith replication yields a $q^{(i)}(p, n, s)$, which is given by

$$q^{(i)}(p, n, s) = \frac{1}{1 + r^{(i)}(p, n, s) \cdot d^{(i)}(p, n, s)}, \quad \forall i \in \mathbb{N}, \tag{4}$$

where, $r^{(i)}(p, n, s)$ is the number of evaluations of the objective function (NEOF) for each ith replication, and $d^{(i)}(p, n, s)$ is the distance to the true point (DTP), which is given by

$$d^{(i)}(p, n, s) = +\sqrt{(\hat{x} - \tilde{x}_i)^t (\hat{x} - \tilde{x}_i)}, \, \forall i \in \mathbb{N}, \tag{5}$$

where: $\hat{x} \in \mathbb{R}^n$ is the theoretical and unique location of the global minimum of the objective function, because we have selected a set of test problems with multimodal objective functions, which each of them has several local minima and just one global minimum; and $\tilde{x}_i \in \mathbb{R}^n$ is the best point reported by the algorithm method $s \in \{\text{GofP}, \text{SGA}, \text{RGA}\}$ at the ith replication.

Obviously, analyzing (4), each ith quality $q^{(i)}(p, n, s)$ sample can uniquely take value in the interval $(0, 1]$, where the best quality of the performance measure is equal to 1, whilst the worst performance causes a $q^{(i)}(p, n, s)$ approaching to zero.

Let $Q(p, n, s)$ be the quality random variable, which is doubtless given by

$$Q(p, n, s) = \frac{1}{1 + R(p, n, s) \cdot D(p, n, s)}, \tag{6}$$

where $R(p, n, s) \in \mathbb{N}_+$ is the NEOF discrete random variable with an unknown distribution, and $D(p, n, s) \in \mathbb{R} \setminus \mathbb{R}_-$ is the DTP continuous random variable, which has also an unidentified distribution.

Notice that, the density function of Q, $f_Q(q) = 0$ for all $q \in \mathbb{R} \setminus (0, 1]$. Moreover, if the optimization method under study yielded values of Q approach to 0, then this should be consider a substandard performance of the method. Whilst if the method produced Q values between 0.9 and 1, we could then say that the method has had an excellent performance.

It is opportune saying that in this comparative study, neither parametric nor nonparametric statistical tests have here been applied, because: firstly, the performance measure $Q(p, n, s)$ has an unknown distribution, although, under some assumptions for both random variables $R(p, n, s)$ and $D(p, n, s)$, it could be easily estimated; secondly, according to a set of preliminary numerical experiments and from its respective empirical frequency histograms, these have shown to be significantly asymmetric, which could implicate that the median and the mean of $Q(p, n, s)$ could have significantly dissimilar values for the metaheuristic methods, it would therefore cause distorted conclusions on the comparison of the optimization methods.

Hence, we have added a set of statistical summaries of $Q(p, n, s)$, $R(p, n, s)$ and $D(p, n, s)$ for each pth problem, which reports: the minimum, the first quartile, the median, the third quartile and the maximum of $Q(p, n, s)$, $R(p, n, s)$ and $D(p, n, s)$. These statistical summaries offer an additional viewpoint in our comparative analysis of the GofP and the GAs.

Furthermore, for carrying out the comparison of the metaheuristic methods, we have considered appropriate to estimate the complementary cumulative frequency function (CCFF) $f\{Q(p, n, s) > q\}$ for all $q \in \mathbb{R}$, through the relative frequency of occurrence of that $\{Q(p, n, s) > q\}$. Namely,

$$f(q) = N_{q+}/N, \quad \forall 0 < q \leq 1, \tag{7}$$

where N_{q+} is the number of ith replication, whose $q^{(i)}(p, n, s)$ have yielded a value more than q, and N is the total number of replications for each pth test n-dimensional problem using the $s \in \{\text{GofP}, \text{SGA}, \text{RGA}\}$ metaheuristic method.

5   Numerical Examples

A set of five numerical test problems defined in 2, 5 and 10 dimensional real field are here presented, where for the set of test problems, the parameters of the GofP methods were fixed as follows: $\alpha = 0.9$, $\delta = 5$, $\varepsilon = 10^{-6}$, the number of sampling or replications of the GofP was equal to 100, the number of players $\eta = 5$, the start point for each ℓth randomized pattern searches $\mathcal{H}_\ell^{[0]}$ was randomly located using a uniform random number between -10 and 10 for each $\nu_{0,\ell}$ component, and the initial balance for each player was fixed according to each n-dimensional case, this is, $b_\ell^{[0]}$ was fixed to 400, 500 and 700; for the case of n equal to 2, 5 and 10, respectively.

Whilst, the parameters of the SGA and the RGA were fixed as follows: for each kth individual of the initial population has been also randomly located, using a uniform random number between -20 and 20 for each ith component of the vector $x_k \in \mathbb{R}^n$ that represents it; the number of members of the initial population, which is denoted by z, was fixed to $50n$; a crossover probability equal to 0.7; a mutation probability of 0.1; a number of couple equal to $55n$; and each ith component of the vector x_k, which represents each kth individual, is bounded between -20 and 20. Namely, $-20 \leq x_k^{(i)} \leq 20$ for all $i \in \{1, \dots, n\}$ and for each member in $\{x_k\}_{k=1}^z$.

The stopping rule for the SGA was $|f_g(\hat{x}) - f_g(\check{x})| < \varepsilon$, where $f_g(\hat{x})$ is the obtained maximum value of $f(x)$ at the gth generation, $f_g(\check{x})$ is the obtained minimum value of $f_g(x)$ at the same generation, and $\varepsilon = 10^{-6}$ is the stopping threshold. Nevertheless, the stopping rule for the RGA was defined by a number of generations equal to 400.

It is worthwhile pointing out that we have run the SGA under fine tuning of the parameters of the GA, which are handled by the SGA itself, and the number of replications for getting samples of solving the problems by the GofP, and the GAs were all equals to 100.

5.1   Goldstein Price Problem, $p = 1$

This problem was considered by Shi and Ólafsson [13], because it is commonly used to test global optimization, and we have also considered this problem for same reason.

$$\underset{x \in \mathbb{R}^2}{\text{minimize}} f(x), \tag{8}$$

$$\text{subject to: } -2.5 \leq x^{(i)} \leq 2.5, \quad \forall i \in \{1, 2\}, \tag{9}$$

where

$$
\begin{aligned}
f(x) =& (1 + (x^{(1)} + x^{(2)} + 1)^2 \\
& (19 - 14x^{(1)} + 3x^{(1)^2} - 140x^{(2)} + 6x^{(1)}x^{(2)} + 3x^{(2)^2})) \\
& (30 + (2x^{(1)} - 3x^{(2)})^2 \\
& (18 - 32x^{(1)} + 12x^{(1)^2} + 48x^{(2)} - 36x^{(1)}x^{(2)} + 27x^{(2)^2})).
\end{aligned}
\tag{10}
$$

For this problem, penalty function approach was used, then for a $k = 10^3$, the problem was then rewritten as

$$\underset{x \in \mathbb{R}^2}{\text{minimize}} \left(f(x) + p(x) \right), \tag{11}$$

where $p(x) = k \sum_{i=1}^{2} \left(\max(-2.5 - x^{(i)}, 0) + \max(x^{(i)} - 2.5, 0) \right)$.

Solution: $\hat{x} = (0, -1)^t$ is the global solution, and with a value of $f(\hat{x}) = 3$.

Table 1 depicts a report of statistical summary for $Q(1, 2, s)$, $D(1, 2, s)$ and $R(1, 2, s)$, which were yielded by the metaheuristic methods. In the table we can see the GofP reached a significative better performance $Q(1, 2, \text{GofP})$ than the yielded values by the GAs. Moreover, from the reported results on the table, we can say the $D(1, 2, \text{GofP})$ was smaller than the values of $D(1, 2, \text{SGA})$ and $D(1, 2, \text{RGA})$. Nevertheless, the GofP required more number of function evaluations than the GAs.

Table 1. Statistical summary of $Q(1, 2, s)$, $D(1, 2, s)$ and $R(1, 2, s)$ by metaheurictic method for the Goldstein Price problem in $\mathbb{R}^2$

	Method	Min	1st Qu.	Median	Mean	3rd Qu.	Max
$Q(1, 2, s)$	GofP	0.999903	0.999956	0.999970	0.999967	0.999984	0.999998
	SGA	0.091025	0.978662	0.985741	0.942320	0.990963	0.998432
	RGA	0.003683	0.017505	0.026627	0.057548	0.053657	0.835909
$D(1, 2, s)$	GofP	0.0e+00	0.000000	0.000000	0.000000	0.000000	0.000000
	SGA	0.0e+00	0.000003	0.000004	0.000072	0.000007	0.001668
	RGA	2.5e−05	0.003923	0.007578	0.014796	0.016221	0.151307
$R(1, 2, s)$	GofP	3522	5480	6431	6309	7298	9485
	SGA	2785	3081	3280	3395	3500	6975
	RGA	1788	3286	4291	4404	5384	8420

5.2 Rastrigin Problem, $p = 2$

This problem is an extension to the n-dimensional Euclidean space $\mathbb{R}^n$ from the original Rastrigin problem, which has been proposed by Mühlenbein and coworkers [9].

$$\underset{x \in \mathbb{R}^n}{\text{minimize}} \left(10n + \sum_{i=1}^{n} \left[x^{(i)^2} - 10 \cos(2\pi x^{(i)}) \right] \right). \tag{12}$$

Solution: $\hat{x} = (\underbrace{0, 0, \ldots, 0}_{n})^t$ is the global solution, and with value function $f(\hat{x}) = 0$.

Table 2. Statistical summary of $Q(2,2,s)$, $D(2,2,s)$ and $R(2,2,s)$ by metaheurictic method for the Rastrigin problem in $\mathbb{R}^2$

	Method	Min	1st Qu.	Median	Mean	3rd Qu.	Max
$Q(2,2,s)$	GofP	0.999845	0.999887	0.999906	0.999910	0.999930	0.999986
	SGA	0.000228	0.971631	0.981521	0.885927	0.988382	0.997171
	RGA	0.003129	0.056990	0.123612	0.162379	0.216709	0.807428
$D(2,2,s)$	GofP	0.0e+00	0.000000	0.000000	0.000000	0.000000	0.000000
	SGA	1.0e−06	0.000003	0.000005	0.049901	0.000009	0.994964
	RGA	4.3e−05	0.000653	0.001652	0.005960	0.004280	0.195209
$R(2,2,s)$	GofP	3220	5334	5758	5648	6188	6939
	SGA	2841	3199	3470	3556	3772	5553
	RGA	1632	3295	4512	4640	5812	8756

Table 3. Statistical summary of $Q(2,5,s)$, $D(2,5,s)$ and $R(2,5,s)$ by metaheurictic method for the Rastrigin problem in $\mathbb{R}^5$

	Method	Min	1st Qu.	Median	Mean	3rd Qu.	Max
$Q(2,5,s)$	GofP	0.993528	0.996719	0.999631	0.998324	0.999710	0.999954
	SGA	0.000024	0.000036	0.000044	0.001045	0.000058	0.013761
	RGA	0.001155	0.002701	0.004026	0.005327	0.007055	0.018407
$D(2,5,s)$	GofP	0.000000	0.000000	0.000000	0.000000	0.000001	0.000001
	SGA	0.004022	0.990782	1.003044	1.072469	1.412059	2.229021
	RGA	0.001709	0.005985	0.011131	0.016601	0.020991	0.097805
$R(2,5,s)$	GofP	4206	5807	7814	7951	10038	12455
	SGA	14219	17044	18798	19043	20417	26009
	RGA	7502	15491	19420	20941	24638	48803

As is shown in Table 2, the GofP reported a more advantageous value of $Q(2,2,\text{GofP})$ than the GAs, because the minimum $Q(2,2,\text{GofP})$ was 0.999845, whilst the minimum values of $Q(2,2,\text{SGA})$ and $Q(2,2,\text{RGA})$ were respectively 0.000228 and 0.003129. Furthermore, the reported mean of $Q(2,2,\text{GofP})$ was 0.999910, in contrast to the means yielded by the SGA and the RGA, which were 0.885927 and 0.162379, respectively. It can be also seen from the table, the $D(2,2,\text{GofP})$ was practically zero, whilst the SGA and RGA respectively reported minimum values of DTP 1.0e−06 and 4.3e−05. In relation to $R(2,2,s)$, there exists a no significant difference among the metaheuristic methods, what does not allow us to distinguish the best metaheuristic method, if we comparatily analysis the performance of them using their $R(2,2,s)$.

On the other hand, according to Table 3, the performance of the GofP reached a considerably better level of the $Q(2,5,s)$ than the $Q(2,5,s)$ of both GAs, because the minimum value of $Q(2,5,\text{GofP})$ was equal to 0.993528, whilst the reported maximum values of $Q(2, 5, \text{SGA})$ and $Q(2, 5, \text{RGA})$ were respectively 0.013761 and 0.018407. Moreover, from the table, it can be seen that the maximum DTP of GofP ($D(2, 5, \text{GofP})$) was 0.000001, in contrast to the reported minimum values by the GAs, which were 0.004022 and 0.001709; and in reference to the $R(2, 5, s)$, the GofP obtained a maximum value of $R(2, 5, \text{GofP})$ equal to 12455, whilst the SGA and RGA respectively reported maximum values of 26009 and 48803.

From Table 4, we have that the GofP had a much higher performance than the GAs, because the minimum value of $Q(2, 10, \text{GofP})$ was equal to 0.975781, whilst the reported maximum values of $Q(2, 10, \text{SGA})$ and $Q(2, 10, \text{RGA})$ were 0.000027 and 0.002115, respectively. Moreover, the maximum of $D(2, 10, \text{GofP})$ was 0.000002; to difference of the minimum values reported by the GAs, which were 0.968809 and 0.005161. On the NEOF, we can say the maximum value of $R(2, 10, \text{GofP})$ was equal to 23033, in contract to the minimum values of the GAs, these were 37516 and 20957.

Table 4. Statistical summary of $Q(2, 10, s)$, $D(2, 10, s)$ and $R(2, 10, s)$ by metaheurictic method for the Rastrigin problem in $\mathbb{R}^{10}$

	Method	Min	1st Qu.	Median	Mean	3rd Qu.	Max
$Q(2, 10, s)$	GofP	0.975781	0.984336	0.987424	0.989437	0.998704	0.999376
	SGA	0.000005	0.000007	0.000008	0.000009	0.000010	0.000027
	RGA	0.000101	0.000511	0.000731	0.000807	0.000984	0.002115
$D(2, 10, s)$	GofP	0.000000	0.000000	0.000001	0.000001	0.000002	0.000002
	SGA	0.968809	1.984582	2.426535	2.455552	2.801321	4.665179
	RGA	0.005161	0.012359	0.017905	0.028672	0.028657	0.222039
$R(2, 10, s)$	GofP	6939	9072	11006	12229	14373	23033
	SGA	37516	46870	49286	49506	52281	67703
	RGA	20957	59961	71744	72416	84468	134774

5.3 Tang Problem, $p = 3$

This problem is presented by Shi and Ólafsson [13].

$$\underset{x \in \mathbb{R}^n}{\text{minimize}} \sum_{i=1}^{n} \left[\sin(x^{(i)}) + \sin(2x^{(i)}/3) \right], \tag{13}$$

$$\text{subject to: } 3 \le x^{(i)} \le 13; \ \forall i \in \{1, \ldots, n\}. \tag{14}$$

For this problem, a penalty function viewpoint was used, the problem is then defined by

$$\underset{x \in \mathbb{R}^2}{\text{minimize}} \left(\sum_{i=1}^{n} \left[\sin(x^{(i)}) + \sin(2x^{(i)}/3) \right] + p(x) \right), \tag{15}$$

where

$$p(x) = 10^6 \sum_{i=1}^{n} \left(\max(3 - x^{(i)}, 0) + \max(x^{(i)} - 13, 0) \right).   \tag{16}$$

Solution: $\hat{x} = \underbrace{(5.362247554154065, \ldots, 5.362247554154065)^t}_{n}$ is the global solution, and with a function value of $f(\hat{v}) = -1.215598217508091n$.

Table 5 depicts that the GofP reached a significantly superior response of $Q(3, 2, s)$ in comparison to the generated performances by the GAs. As we can see from table, the reported interval of the GofP between the first quartile and third quartile of $Q(3, 2, \text{GofP})$ was $(0.999891, 0.999958)$. However, the intervals yielded by the SGA and the RGA were respectively equals to $(0.715244, 0.854145)$ and $(0.012875, 0.025398)$, what allows us to verify the good performance reached by the GofP in comparison to the GAs in this case. Moreover, the reported maximum values of $D(3, 2, \text{GofP})$ was zero, whilst the SGA and RGA reported maximum values of 0.000521 and 0.153334, respectively.

Table 5. Statistical summary of $Q(3, 2, s)$, $D(3, 2, s)$ and $R(3, 2, s)$ by metaheurictic method for the Tang problem in $\mathbb{R}^2$

	Method	Min	1st Qu.	Median	Mean	3rd Qu.	Max
$Q(3, 2, s)$	GofP	0.999108	0.999891	0.999927	0.999916	0.999958	0.999997
	SGA	0.447584	0.715244	0.773974	0.780454	0.854145	0.995261
	RGA	0.003549	0.012875	0.016408	0.021733	0.025398	0.094698
$D(3, 2, s)$	GofP	0.000000	0.000000	0.000000	0.000000	0.000000	0.000000
	SGA	0.000002	0.000072	0.000117	0.000124	0.000159	0.000521
	RGA	0.002282	0.009803	0.012311	0.015311	0.014742	0.153334
$R(3, 2, s)$	GofP	2569	4719	5326	5292	6048	7147
	SGA	1964	2303	2366	2470	2504	4373
	RGA	1831	3595	4348	4569	5384	11346

From Table 6, we can see that the performance of the GofP was significantly better than the performance of the GAs, because the interval of $Q(3, 5, s)$ between its first quartile and third quartile was $(0.995418, 0.998098)$, whilst the reported intervals of the SGA and the RGA were respectively $(0.003078, 0.008684)$ and $(0.000925, 0.002008)$. We also note that the mean of $D(3, 5, s)$ for the GofP, SGA and RGA were 0.865507, 0.214862 and 0.063269, respectively. Furthermore, note from the table that, the GofP required a maximum $R(3, 5, \text{GofP})$ significantly smaller that the GAs.

Table 6. Statistical summary of $Q(3,5,s)$, $D(3,5,s)$ and $R(3,5,s)$ by metaheurictic method for the Tang problem in $\mathbb{R}^5$

	Method	Min	1st Qu.	Median	Mean	3rd Qu.	Max
$Q(3,5,s)$	GofP	1.5e−05	0.995418	0.996548	0.827801	0.998098	0.999872
	SGA	7.0e−06	0.003078	0.004486	0.012726	0.008684	0.221331
	RGA	2.9e−04	0.000925	0.001386	0.001605	0.002008	0.008778
$D(3,5,s)$	GofP	0.000000	0.000000	0.000001	0.865507	0.000001	5.091215
	SGA	0.000356	0.006932	0.012296	0.214862	0.017641	5.101781
	RGA	0.008304	0.020350	0.034852	0.063269	0.076490	0.495136
$R(3,5,s)$	GofP	4051	4492	5042	6259	7702	13191
	SGA	9870	14374	16744	17796	20803	28740
	RGA	6099	12136	19862	20459	26251	55217

As we can be seen from Table 7, the $Q(3,10,s)$ mean obtained by the GofP, the SGA and the RGA were respectively 0.197418, 0.000369 and 0.000297. However, in this case, the GAs got values of $D(3,10,s)$ better than the GofP. With respect to the NEOF, the GofP required a significative smaller number of $R(3,10,\texttt{GofP})$ than the GAs.

5.4 W Problem, $p = 4$

This problem is here proposed by the author, which is given by

$$\underset{x \in \mathbb{R}^n}{\text{minimize}} \sum_{i=1}^{n} \left[\left(\frac{x^{(i)}}{4} \right)^4 + 1 - (x^{(i)} - 2)^2 \right]. \tag{17}$$

Notice that, the objective function for optimizing is multimodal function and continuous in all Euclidean space $\mathbb{R}^n$, and it only has a global minimum.

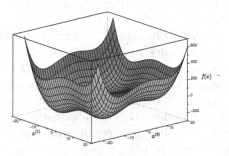

Fig. 1. W function

Table 7. Statistical summary of $Q(3, 10, s)$, $D(3, 10, s)$ and $R(3, 10, s)$ by metaheurictic method for the Tang problem in $\mathbb{R}^{10}$

	Method	Min	1st Qu.	Median	Mean	3rd Qu.	Max
$Q(3, 10, s)$	GofP	8.0e−06	0.000017	0.000022	0.197418	0.000026	0.992883
	SGA	3.0e−06	0.000279	0.000343	0.000369	0.000437	0.000826
	RGA	7.6e−05	0.000225	0.000291	0.000297	0.000358	0.000680
$D(3, 10, s)$	GofP	0.000001	5.091214	5.091215	4.643824	5.091217	8.818244
	SGA	0.028145	0.043804	0.052133	0.250084	0.058609	5.044742
	RGA	0.020157	0.035721	0.051613	0.095453	0.084864	0.856772
$R(3, 10, s)$	GofP	6841	8064	8832	9484	9819	20145
	SGA	35038	47478	56661	57278	65694	96636
	RGA	13618	52792	66440	65730	80356	116950

Table 8. Statistical summary of $Q(4, 2, s)$, $D(4, 2, s)$ and $R(4, 2, s)$ by metaheurictic method for the W function problem in $\mathbb{R}^2$

	Method	Min	1st Qu.	Median	Mean	3rd Qu.	Max
$Q(4, 2, s)$	GofP	0.000007	0.999127	0.999396	0.919438	0.999538	0.999934
	SGA	0.061023	0.825318	0.884586	0.836627	0.929947	0.988410
	RGA	0.001911	0.011449	0.030554	0.059176	0.067604	0.424724
$D(4, 2, s)$	GofP	0.000000	0.000000	0.000000	1.787342	0.000000	22.341778
	SGA	0.000004	0.000028	0.000047	0.000127	0.000072	0.002498
	RGA	0.000177	0.002280	0.005932	0.024284	0.021248	0.278545
$R(4, 2, s)$	GofP	2456	2597	4379	4300	5655	7193
	SGA	2126	2490	2612	3106	2746	10277
	RGA	1605	3334	4883	5057	6374	11690

Figure 1 depicts an example of the W function in $\mathbb{R}^2$, which shows four minima and only one of them is a global minimum, that is located at the point $\hat{x} = (-12.2055, -12.2055)^t$.

Solution: $\hat{x} = (-12.20549696692415, \ldots, -12.20549696692415)^t$ is the global solution, and with minimum value function $f(\hat{x}) = -114.10356900557n$.

From Table 8, we can note that both the GofP and the SGA reported similar results of $Q(4, 2, s)$, whilst the $Q(4, 2, \text{RGA})$ was substantially worse than the yielded by the GofP and the SGA. However, the obtained value of $D(4, 2, \text{SGA})$ was considerably smaller than the others metaheuristic methods. It is worthwhile pointing out that the GofP reaches these results with a maximum of $R(4, 2, \text{GofP})$ equal to 7193, in contrast with the GAs, which reported maximum value of 10277 and 11690.

As we can be seen from Table 9, the GofP method has shown to have higher performance than the GAs, because the mean of $Q(4, 5, \text{GofP})$ was 0.029873,

Table 9. Statistical summary of $Q(4, 5, s)$, $D(4, 5, s)$ and $R(4, 5, s)$ by metaheuristic method for the W function problem in $\mathbb{R}^5$

	Method	Min	1st Qu.	Median	Mean	3rd Qu.	Max
$Q(4, 5, s)$	GofP	2.0e−06	0.000004	0.000007	0.029873	0.000010	0.996253
	SGA	1.0e−06	0.000734	0.001239	0.003528	0.002510	0.156065
	RGA	5.9e−05	0.000102	0.000119	0.000132	0.000159	0.000312
$D(4, 5, s)$	GofP	0.000001	22.341778	31.596045	28.925895	31.596045	44.683556
	SGA	0.000183	0.009940	0.014141	1.800320	0.021206	22.379268
	RGA	0.086110	0.208527	0.314682	0.391693	0.496187	1.633537
$R(4, 5, s)$	GofP	3813	4044	4353	5773	6938	12356
	SGA	14766	35887	57028	55623	74738	119282
	RGA	5176	18012	27302	27857	36281	60624

Table 10. Statistical summary of $Q(4, 10, s)$, $D(4, 10, s)$ and $R(4, 10, s)$ by metaheuristic method for the W function problem in $\mathbb{R}^{10}$

	Method	Min	1st Qu.	Median	Mean	3rd Qu.	Max
$Q(4, 10, s)$	GofP	1e−06	4.0e−06	5.0e−06	0.079089	6.0e−06	0.991331
	SGA	0e+00	0.0e+00	0.0e+00	0.000016	3.3e−05	0.000084
	RGA	8e−06	1.1e−05	1.2e−05	0.000013	1.4e−05	0.000018
$D(4, 10, s)$	GofP	0.000001	22.341778	22.341779	25.799132	31.59605	38.697095
	SGA	0.037815	0.069033	22.297555	18.332901	22.36464	31.625469
	RGA	0.419649	0.533760	0.663045	0.810854	0.90135	5.887933
$R(4, 10, s)$	GofP	6664	6989	7322	8393	8816	18944
	SGA	100570	233857	273234	290789	337276	554248
	RGA	16538	94410	119361	115935	145801	146657

whilst the reported means by the SGA and the RGA were 0.003528 and 0.000119, respectively. Nevertheless, with respect to the $D(4, 5, s)$, the GAs obtained significative better results than the GofP method, because the means were respectively 1.800320 and 0.391693 for the SGA and the RGA, whilst the mean of $D(4, 5, \mathtt{GofP})$ was 28.925895. Note that, the $R(4, 5, \mathtt{GofP})$ was considerably smaller the reported values by the GAs.

From Table 10, we note that, the GofP has shown to have a better performance $Q(4, 10, \mathtt{GofP})$ than the reported performance by the GAs. Nevertheless, the GofP and the SGA yielded maximum value of $D(4, 10, s)$ greater than 30, although the reported NEOF of GofP was significatively smaller than the registered value of the GAs.

Figure 2 displays a set of figures, which represent the complementary cumulative frequency of $Q(p, 2, s)$ for $p \in \{1, 2, 3, 4\}$ and $s \in \{\mathtt{GofP}, \mathtt{SGA}, \mathtt{RGA}\}$. As we can be seen from figures, the GofP yielded a considerably superior performance of $Q(p, 2, \mathtt{GofP})$ for $p \in \{1, 2, 3\}$ than the GAs.

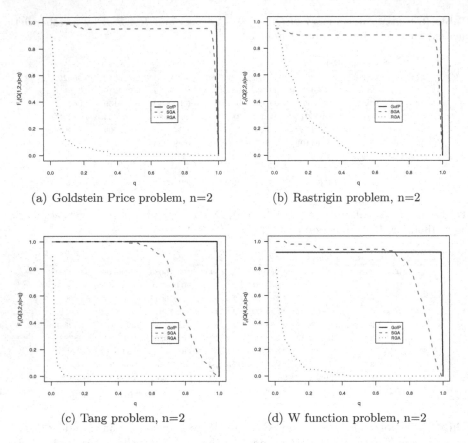

(a) Goldstein Price problem, n=2   (b) Rastrigin problem, n=2

(c) Tang problem, n=2   (d) W function problem, n=2

Fig. 2. Complementary cumulative frequency of quality performance $Q(p, 2, s)$ for a set of problems in $\mathbb{R}^2$

6   Conclusions and Future Research

This article has provided a numerical comparison of the GofP and two implementations of the GAs. Although this comparative study is based on a small set of numerical examples, we think that the GofP has offered better performance than the two implementations of the GAs, what it would allow us to infer that the GofP could be a good via for globally solving nonlinear optimization problems.

However, thus far, that the GofP has not tuned and a stopping rule based on the convergence rate has not been included yet, which will require to do more study for tuning it and improving its stopping rule.

We shall end with some interesting future research, which would allow us to focus on: (i) to estimate the convergence rate of the GofP; (ii) to tune the parameters of the GofP; (iii) to propose of an optimum payment rule for dealing points among the active players at each game round; and (iv) to study a very large number of numerical examples for statistically evaluating the GofP.

References

1. Bajpai, P., Kumar, M.: Genetic algorithm - an approach to solve global optimization problems. Indian J. Comput. Sci. Eng. **1**(3), 199–206 (2010)
2. Baudin, M., Couvert, V., Steer, S.: Optimization in Scilab. Consortium Scilab and the National Institute for Research in Computer Science and Control, Le Chesnay Cedex, France (2010)
3. Brea, E.: On the performance of the mixed integer randomized pattern search algorithm. In: González, Y., et al. (eds.) The 13th International Congress on Numerical Methods in Engineering and Applied Sciences, Caracas, vol. 1, pp. Op 61–Op 72 (2016)
4. Brea, E.: Game of patterns: an approach for solving mixed integer nonlinear optimization problems. In: International Congress on Industrial Engineering and Operations Management (IEOM 2017), Bogota, Colombia (2017)
5. Brea, E.: On the game of patterns: a metaheuristic method for globally solving mixed integer nonlinear problems. Metaheuristics (2018). Submitted on 30 April 2018
6. Holland, J.H.: Adaptation in natural and artificial systems : an introductory analysis with applications to biology, control, and artificial intelligence. University of Michigan Press, Ann Arbor (1975)
7. Mazalov, V.V.: Mathematical Game Theory and Applications, 1st edn. Wiley, Chichester (2014)
8. Mitchell, M.: An Introduction to Genetic Algorithms. Complex Adaptive Systems. MIT Press, Cambridge (1996)
9. Muhlenbein, H., Schomisch, M., Born, J.: The parallel genetic algorithm as function optimizer. Parallel Comput. **17**(6–7), 619–632 (1991)
10. R Core Team: R: A Language and Environment for Statistical Computing (2018)
11. Scilab Consortium: Scilab, January 2018
12. Scrucca, L.: GA: a package for genetic algorithms in R. J. Stat. Softw. **53**(4), 1–37 (2013)
13. Shi, L., Ólafsson, S.: Nested partitions method for global optimization. Oper. Res. **48**(3), 390–407 (2000)
14. Whitley, D.: A genetic algorithm tutorial. Stat. Comput. **4**(2), 65–85 (1994)

Stochastic Local Search Algorithms for the Direct Aperture Optimisation Problem in IMRT

Leslie Pérez Cáceres[✉], Ignacio Araya, Denisse Soto,
and Guillermo Cabrera-Guerrero

Pontificia Universidad Católica de Valparaíso, Valparaíso, Chile
{leslie.perez,ignacio.araya,guillermo.cabrera}@pucv.cl
denisse.soto.s@mail.pucv.cl

Abstract. In this paper, two heuristic algorithms are proposed to solve the direct aperture optimisation problem (DAO) in radiation therapy for cancer treatment. In the DAO problem, the goal is to find a set of deliverable aperture shapes and intensities so we can irradiate the tumor according to a medical prescription without producing any harm to the surrounding healthy tissues. Unlike the traditional two-step approach used in intensity modulated radiation therapy (IMRT) where the intensities are computed and then the apertures shapes are determined by solving a sequencing problem, in the DAO problem, constraints associated to the number of deliverable aperture shapes as well as physical constraints are taken into account during the intensities optimisation process. Thus, we do not longer need any leaves sequencing procedure after solving the DAO problem. We try our heuristic algorithms on a prostate case and compare the obtained treatment plan to the one obtained using the traditional two-step approach. Results show that our algorithms are able to find treatment plans that are very competitive when considering the number of deliverable aperture shapes.

Keywords: Intensity modulated radiation therapy
Direct Aperture Optimisation · Multi-leaf collimator sequencing

1 Introduction

Intensity modulated radiation therapy (IMRT) is one of the most common techniques in radiation therapy for cancer treatment. Unfortunately, the IMRT planning problem is an extremely complex optimisation problem. Because of that, the problem is usually divided into three sequential sub-problems, namely, beam angle optimisation (BAO), fluence map optimisation (FMO) and multi-leaf collimator (MLC) sequencing. In the BAO problem, the aim is to find the optimal beam angle configuration (BAC), i.e. the BAC that leads to the optimal treatment plan. To find this optimal treatment plan, the intensities for the given

© Springer Nature Switzerland AG 2019
M. J. Blesa Aguilera et al. (Eds.): HM 2019, LNCS 11299, pp. 108–123, 2019.
https://doi.org/10.1007/978-3-030-05983-5_8

BAC, need to be computed (FMO problem). Finally, the MCL sequencing problem must be solved to find the set of deliverable aperture shapes and intensities of the MLC during the delivery process [10].

One problem of this sequential approach is that, once the FMO problem is solved, i.e., we have computed the optimal intensities for a given BAC, we need to determine a set of apertures for the MLC such that the optimal intensities can be delivered to the patient. Although there exist exact approaches that minimise the time a patient is exposed to the radiation (*beam-on time*) and heuristic algorithms that are very efficient minimising the number of apertures (*decomposition time*), solutions of the MLC sequencing problem usually require too many deliverable aperture shapes and long beam-on times, which is something that we want to avoid as it means patients should stay longer on the treatment couch and fewer patients can be treated per day. In order to reduce the number of apertures and the beam-on time, treatment planners usually simplify the treatment plan by rounding the intensities at each beam angle to some predefined values. Although this (over-)simplification of the optimal plan allows us to deliver a treatment plan using fewer deliverable aperture shapes and shorter beam-on times, the quality of the treatment plan is impaired w.r.t. the optimal one. Thus, it seems reasonable to incorporate some MLC sequencing considerations into the FMO problem such that the optimal intensities found during its optimisation process can be directly delivered to the patient without needing any 'adjustment' process.

In this paper we focus on solving the *direct aperture optimisation* problem (DAO), that is, the problem of optimising the intensities and shapes of the apertures simultaneously, for a specific BAC. In other words, we aim to solve the FMO problem taking into account a constraint on the number of deliverable apertures and the physical constraints associated with the MLC sequencing. In this way, any post-processing on the intensities found by the solver is needed.

First, let us introduce the mathematical model of the FMO problem [3]. Consider that each beam angle is associated with a series of sub-beams or *beamlets* and n is the total number of beamlets summed over all the possible beam angles. Let $\mathscr{A}$ be a BAC and $x \in \mathbb{R}^n$ be an intensity vector or *fluence map* solution. Each component x_i of a solution represents the length of time that a patient is exposed to the *beamlet* i. The set $\mathcal{X}(\mathscr{A}) \subseteq \mathbb{R}^n$ is the set of all feasible solutions of the FMO problem when the BAC $\mathscr{A}$ is considered[1]. Finally, $z(x) : \mathbb{R}^n \to \mathbb{R}_+^* := \{v \in \mathbb{R} : v \geq 0\} \cup \{\infty\}$ is a function that attempts to penalize both: (1) zones where the tumor would not be properly irradiated according to the medical prescription and (2) healthy tissues that would be harmed by the fluence map x. Thus, solving the FMO problem for a given BAC $\mathscr{A}$ consists on finding the fluence map $x \in \mathcal{X}(\mathscr{A})$ that minimises the penalty $z(x)$.

The main difference of DAO w.r.t. the FMO problem is related to the set of feasible intensity vectors $\mathcal{X}(\mathscr{A})$. For the DAO problem we need to add some additional constraints to the set $\mathcal{X}(\mathscr{A})$. In particular, we first need to set the

---

[1] i.e., only beamlets x_i that belong to a beam angle in $\mathscr{A}$ are allowed to be greater than zero.

number of apertures $\Theta_{\mathscr{A}_i}$ for each beam $\mathscr{A}_i \in \mathscr{A}$ with $i = \{1, \ldots, N\}$. Then, we have to ensure that the beamlets intensities can be delivered given the number of apertures for each beam angle $\theta^{\mathscr{A}_i}$. We denote this new set of feasible intensity vectors as $\mathcal{X}(\mathscr{A}, \Theta)$.

The DAO problem was first introduced by [13]. In their paper, the authors identify as input for the DAO problem the beam angles, the beam energies and the number of apertures per beam angle, while the decision variables are the leaf positions for each aperture and the weight assigned to each aperture. They propose to solve the problem by adjusting the leaf position at each iteration following a set of rules that determine, using a probabilistic function, which pair of leaves should be adjusted next. Authors consider a function similar to the one we use in this paper. The DAO problem is also considered in [15]. Unlike previous approaches, [15] proposes a "rapid" DAO algorithm which replaces the traditional Monte Carlo method for dose calculation by what they call a dose influence matrix based piece-wise aperture dose model. The authors claim that their approach is faster than traditional approaches based on Monte Carlo as well as able to find treatment plans that result in more precise dose conformality. In [2], authors focus on the DAO problem as a mean of reducing the complexity of IMRT. Authors define the complexity of a treatment plan in IMRT in terms of the number of *monitor units* (MU) a plan needs to be delivered. Since the DAO problem gives control over the number of apertures (and therefore over the number of MUs), authors claim that they can reduce the complexity of treatment plans without any major impairment on the overall quality of the delivered treatment plan. A similar conclusion is drawn by [8,11,13,15].

The main goal of our work is to study and evaluate the design and application of hybrid algorithms combining the best features of local search techniques and FMO models to generate high-quality treatment plans that exhibit characteristics that make them candidates to be applied in real cases. As a first step towards this goal, the work presented in this paper studies the application of two simple stochastic local search algorithms to solve the DAO problem. We apply these algorithms on a simple prostate case considering two organs at risk, namely, the bladder and the rectum. We compare treatment plans obtained by our algorithms with those obtained by well-known FMO models. We compare these treatment plans in terms of their corresponding dose-volume-histograms as well as in terms of their objective function value in order to evaluate their quality and their practical desirability.

The remaining of this paper is organized as follows: Sect. 1.1 introduces the general concepts of IMRT as well as the mathematical models we will consider in this study. In Sect. 2 the algorithms we implement in this paper are presented and their main features discussed. Section 3 presents the prostate case algorithms are applied on and the obtained results. A discussion on these results is also included in this section. Finally, in Sect. 4 we draw the main conclusions of our work as well as outline future work.

1.1   IMRT: An Overview

In this section we briefly introduce some key concept in IMRT. This section is mainly based on the IMRT description given in [3–6,10].

In IMRT, each organ is discretised into small sub-volumes called *voxels*. The radiation dose deposited by a fluence map x into each voxel j, denoted by d_j^r, of the tumor and each organ at risk (OAR), is calculated using expression (1) [10].

$$d_j^r(x) = \sum_{i=1}^{n} A_{ji}^r x_i \quad \forall j = 1, 2, \ldots, m^r, \tag{1}$$

where $r \in R = \{O_1, \ldots, O_Q, T\}$ is an element of the index set of regions, with the tumor indexed by $r = T$ and the organs at risk and normal tissue indexed by $r = O_q$ with $q = 1, \ldots, Q$. m^r is the total number of voxels in region r, j corresponds to a specific voxel in region r, $d^r \in \mathbb{R}^{m^r}$ is a dose vector and its elements d_j^r give the total dose delivered to voxel j in region r by the fluence map $x \in X(\mathscr{A}, \theta)$. Here, dose deposition matrix $A^r \in \mathbb{R}^{m^r \times n}$ is a given matrix where $A_{ji}^r \geq 0$ defines the rate at which radiation dose along beamlet i is deposited into voxel j in region r.

Based on the dose distribution in (1), both physical (*dose-volume*) and biological (*dose-response*) models have been proposed (see [10] for a survey). Here, the following dose-volume model of the FMO problem is considered:

$$\min_{x \in X(\mathscr{A}, \Theta)} z(x) = \sum_{q=1}^{Q} (\frac{1}{m^{O_q}} \times \sum_{j=1}^{m^{O_q}} (\max(d_j^{O_q} - D^{O_q}, 0)^2)) + \tag{2}$$

$$(\frac{1}{m^T} \times \sum_{j=1}^{m^T} (\max(D^T - d_j^T, 0)^2))$$

where parameters D^T and D^{O_q} correspond to the prescribed dose values for tumor and OARs, respectively. This problem is a convex optimisation problem and, thus, optimal fluence maps can be obtained using mathematical programming techniques.

2   Stochastic Local Search Algorithms for DAO

In this section we present two stochastic local search algorithms to solve DAO. For a given BAC $\mathscr{A}$, and subject to a maximum number of apertures for each beam angle, the algorithms attempt to find a set of aperture shapes for each angle in $\mathscr{A}$ such that the corresponding fluence map x minimizes $z(x)$. Both algorithms follow an iterative improvement scheme and they diverge in the representation they use to search the aperture shapes and intensities.

The algorithmic scheme followed by the proposed techniques is outlined in Algorithm 1. The algorithm receives as input the maximum number of apertures per beam angle (ASIZE), the number of target beamlets (BSIZE), the number of target voxels (VSIZE), the perturbation size (PSIZE), the number of steps without improvement to perform a perturbation (N_RESTART), the aperture pattern initial setup (AP_SETUP), the budget available for the search (BUDGET), and other parameters related to the search strategy instantiated in the algorithm (...).

The idea of the algorithm is simple. An initial configuration or plan S of aperture shapes and intensities for each beam angle is generated (line 2). x_S denotes the fluence map generated by the plan S. Then the algorithm identifies one of the BSIZE beamlets b impacting the most to a subset of the VSIZE *worst* voxels of the organ (resp. tumor) (line 8), i.e., a subset of the voxels with the largest (resp. smallest) deposited doses of radiation. Then, we attempt to improve the current plan by modifying the apertures and intensities related to the selected beamlet (line 9). If the change improves the quality of S, then it is accepted, otherwise we try with another beamlet and the process is repeated. When all selected beamlet modifications have been evaluated without finding an improvement to the current plan or when N_RESTART iterations without an improvement have been reached, a perturbation of size PSIZE is applied (line 7 and line 17). Finally, the best plan found is returned.

```
 1  IterativeImprovement (ASIZE,BSIZE, VSIZE,PSIZE, N_RESTART,AP_SETUP,
      BUDGET, ...); out: Sbest
 2  S ← initializeangles (AP_SETUP);
 3  Sbest ← S;
 4  no_improvement ← 0;
 5  while !termination(BUDGET) do
 6      while improvementExhauted(BSIZE,VSIZE,S) do
 7          ⌊ S ← perturbation(S, PSIZE);
 8      (b, angle) ← select_promising_beamlet (S, BSIZE, VSIZE);
 9      S′ ← search(b, angle, S, ...);
10      no_improvement ← no_improvement + 1;
11      if z(x_{S′}) < z(x_S) then
12          ⌈ no_improvement ← 0;
13          ⌊ S ← S′;
14      if z(x_S) < z(x_{Sbest}) then
15          ⌊ Sbest ← S;
16      if no_improvement ≥ N_RESTART then
17          ⌊ S ← perturbation(S, PSIZE) ;
18  return Sbest
```

Algorithm 1. Iterated improvement outline of the proposed techniques.

2.1   Selecting a Promising Beamlet

The procedure select_promising_beamlet(S, BSIZE,VSIZE) returns one of the BSIZE beamlets impacting the most to a subset of VSIZE voxels of the organs/tumor. In the following, we give details about how these beamlets (and voxels) are selected.

Consider a current fluence map x. From Eqs. (1) and (2) we can deduce that the change in the evaluation of $z(x)$ provided by increasing or decreasing the intensity of a beamlet i in 1 is approximately:

$$\Delta_z(i) = \sum_{(j,r) \in V} A_{ji}^r \frac{\partial z}{\partial d_j^r}(x)$$

where V is the set of all the pairs (voxel, organ) of the problem. Thus, we consider that the most promising beamlets are those beamlets maximizing $|\Delta_z(i)|$. However, given that computing $\Delta_z(i)$ is expensive, mainly due to $|V|$ is very large (of the order of tens of thousands), we only consider a subset of VSIZE representative voxels.

We rank the voxels according to how much the objective function changes if we increase (or decrease) the radiation dose deposited in the voxel. The rate of change is given by $\left| \frac{\partial z}{\partial d_j^r}(x) \right|$, and the voxels are kept sorted by this value in a set V. The procedure select_promising_beamlet then uses the first VSIZE voxels from V for identifying the BSIZE beamlets maximizing $|\Delta_z(i)|$. Finally, one of these beamlets is randomly selected and returned by the procedure.

2.2   Representation of the DAO Search Space

We propose two different approaches for representing the search space of the problem. An *intensity-based representation* which defines a matrix of intensities for each angle maintaining the constraints related to the apertures. And an *aperture-based representation* which defines a set of aperture shapes for each angle and assigns an intensity to each of these shapes. In the following we provide more details for each representation.

Intensity-Based Representation: The set of aperture shapes of each angle is represented by a single intensity integer matrix I. Each value I_{xy} in the matrix corresponds to the total intensity delivered by the corresponding beamlet and the set of apertures. In order to respect the limit of allowed aperture shapes we force the matrix to respect two conditions:

– The number of different intensities n in the matrix cannot be greater than the maximum number of allowed aperture shapes.
– For each row, consider that the k-th beamlet has the greatest intensity. Then, the intensities of beamlets $\{1, 2, .., k\}$ should increase monotonically and the intensities of beamlets $\{k, k+1, .., x_{max}\}$ should decrease monotonically.

If the intensity matrix I satisfies the two conditions, it is easy to map I to a set of n aperture shapes. Consider the intensity matrix of Fig. 1. It has $n = 5$ different intensities: $\mathcal{Y} = \{1, 2, 4, 7, 8\}$ and each row satisfies the second condition. In order to obtain the intensities of the matrix we can consider the following five aperture shapes:

- An aperture of intensity 1 considering every beamlet (x, y) such that $I_{xy} \geq 1$.
- An aperture of intensity 1 considering every beamlet (x, y) such that $I_{xy} \geq 2$.
- An aperture of intensity 2 considering every beamlet (x, y) such that $I_{xy} \geq 4$.
- An aperture of intensity 3 considering every beamlet (x, y) such that $I_{xy} \geq 7$.
- An aperture of intensity 1 considering every beamlet (x, y) such that $I_{xy} \geq 8$.

Note that, in general, we should consider every aperture with intensity $\mathcal{Y}_i - \mathcal{Y}_{i-1}$ and beamlets (x, y) such that $I_{xy} \geq \mathcal{Y}_i$. Also note that the sum of intensities of the apertures (beam-on time) is equal to the maximum intensity in the matrix $(1 + 1 + 2 + 3 + 1 = 8$ in the example).

Fig. 1. An intensity matrix representing a set of five aperture shapes.

Aperture-Based Representation: We directly represent the aperture shapes of an angle as a set of n apertures $A = \{a_1, \ldots, a_n\}$, where each element is a list $a_i = \{(x_1, y_1), \ldots, (x_r, y_r)\}$, $x_j, y_j \in \{1, \ldots, c_j\}$, and $x_j \leq y_j, \forall j \in \{1, \ldots, r\}$, r is the number of rows in a beam of the MLC, and c_j is the number of active beamlets in row j. The set a indicates the range of beamlets that are open per each row of the beam. An intensity value $I[i] \in \{1, \ldots, \max\_i\}$ is assigned to each of these aperture shapes, where $\max\_i$ is the maximum intensity allowed for a beam/aperture. All open beamlets in an aperture a_i emit radiation with the same intensity $I[i]$. Figure 2 shows an example of this representation using 5 aperture shapes for the matrix presented in Fig. 1. Searching this representation has the benefit that the generated solutions inherently satisfy the constraints associated to the number of apertures. Nevertheless, the complex dependencies of the aperture shapes and the delivered radiation dose difficult the optimisation and thus, effectively searching the solution space depends greatly of the technique applied.

2.3 Intensity-Based Beamlet Targeted Search

When the intensity-based representation is used, the search method of the algorithm attempts to increase (or decrease) the intensity of a subset of contiguous beamlets of the angle. Algorithm 2 shows the procedure. Note that, in addition to the angle, the beamlet and the current solution, the procedure also gets some user-defined parameters as input. First, the values *delta_intensity*

```
a1 = {( 3,10), ( 6,13), (4, 4), ( 1, 5), (1,1)}, I[1]=1
a2 = {( 5, 9), ( 7,13), (6,10), ( 6,14), (1,3)}, I[2]=2
a3 = {( 5, 9), ( 9,13), (5,10), (13,14), (1,2)}, I[3]=4
a4 = {(10,11), (13,13), (6,11), (14,14), (1,6)}, I[4]=1
a5 = {( 8, 8), ( 7, 8), (9,10), (13,13), (3,5)}, I[5]=1
```

Fig. 2. Aperture-based representation for angle intensity matrix in Fig 1.

and a are randomly selected from an uniform distribution). Then, the procedure increaseIntensity attempts to increase in *delta_intensity* the intensity of a square $a \times a$, including the beamlet, and its surrounding beamlets. Note that if $a = 1$ the intensity of only one beamlet will be modified.

Note that the values of the parameters MAX_A and MAX_D progressively decrease as the search goes on (providing that $\alpha < 1$). This mechanism forces larger changes at the beginning of the search and smaller changes at the end. The procedure returns the new modified solution S'.

---

1 search $(b, angle, S, \alpha, \text{MAX\_D}, \text{MAX\_R},)$; **out:** S'
2 *delta_intensity* $\leftarrow$ random(1,MAX_D);
3 MAX_D $\leftarrow$ $\alpha*$MAX_D;
4 $a \leftarrow$ random(1,MAX_R);
5 MAX_R $\leftarrow$ $\alpha*$MAX_R;
6 $S' \leftarrow$ increaseIntensity$(S, b, angle, delta\_intensity, a)$;
7 **return** S';

Algorithm 2. Search procedure for the intensity-based representation.

Increasing/Decreasing Intensities in the Matrix: In Algorithm 2, the procedure increaseIntensity *attempts* to increase the intensity of a square of beamlets in *delta_intensity* units. The procedure increases the intensities directly in the intensity matrix and then it runs a two-phase reparation mechanism for re-satisfying the conditions of the intensity-based representation.

In the first phase we perform a reparation related to the second condition. The procedure repairs each row independently. If *delta_intensity* is positive we may arrive to a situation as in Fig. 3-left. The figure shows the intensities (y axis) of all the beamlets of a row (x axis). The intensity of red beamlets has increased violating the second condition. For repairing the row we first identify the beamlet with the largest intensity b. Then we *increase* the intensities of some beamlets (green beamlets in Fig. 3-right) to force the intensities on the left side of b to increase monotonically and the intensities on the right side of b to decrease monotonically. As a result we obtain the intensity distribution in Fig. 3-right.

In the case that *delta_intensity* is negative we also identify the beamlet b with the largest intensity and force the intensities of the left and the right side of b to increase or decrease monotonically depending on the case. However, in this case this is achieved by *reducing* the intensities of some beamlets.

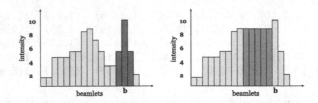

Fig. 3. Example of the reparation mechanism when the intensity of some beamlets *increases*. (left) The intensities in red increased. (right) The reparation mechanism modified the intensities in green. (Color figure online)

In the second phase of the reparation mechanism, called *aperture-reduction*, the first condition is treated. If the number of different intensity values is larger than the maximal number of aperture shapes allowed then we try to reduce this number with the following procedure:

- We consider the set of different intensities $\mathcal{Y}$, where $\mathcal{Y}_0 = 0$.
- We consider that changing an intensity I_{xy} in c units has a cost of c.
- For each value $\mathcal{Y}_i$ ($i \geq 1$), we compute the cost of changing all the intensities $I_{xy} = \mathcal{Y}_i$ by $\mathcal{Y}_{i-1}$ and by $\mathcal{Y}_{i+1}$.
- We perform the change with the minimum cost.
- The process is repeated until the number of different intensities is equal to the number of maximal number of apertures allowed.

Consider the intensity matrix of Fig. 1 and suppose that the maximal number of apertures allowed is 4. The set of intensities is $\mathcal{Y} = \{0, 1, 2, 4, 7, 8\}$. There are 11 beamlets with intensity equal to 1, thus changing its intensities to 0 (or 2) has a cost of 11. Changing the 2s to 1s has a cost of 10. Changing the 4s to 2s has a cost of $4 \cdot 2 = 8$. Changing the 7s to 8s has a cost of 14 and changing the 8s to 7s has a cost of 5. Thus, we will prefer to perform the last change transforming the matrix into the intensity matrix of Fig. 4.

0	0	1	1	7	7	7	7	7	2	1	0	0	0
0	0	0	0	0	1	4	4	7	7	7	7	7	0
0	0	0	1	4	7	7	7	7	7	1	0	0	0
1	1	1	1	1	2	2	2	2	2	2	2	7	7
7	7	4	2	2	1	0	0	0	0	0	0	0	0

Fig. 4. The matrix of Fig. 1 reduced to a set of four aperture shapes.

Perturbation: In the case of the intensity-based representation, the perturbation method performs PSIZE times the following two steps:

- Select a random angle

– Reduce the number of intensities in the angle matrix by performing the aperture-reduction procedure.

This method has a two-fold objective: one is to perturb the current solution in order to explore other regions in the search space, the other to reduce the number of apertures without affecting too much the cost of the objective function.

2.4 Aperture-Based Beamlet Targeted Search

When searching the aperture-based representation the proposed algorithm attempts to either increase or decrease the contribution of radiation of the targeted beamlet by (1) opening/closing the beamlet in the apertures of the angle, or (2) by modifying the intensity of the apertures in which the beamlet is active (not closed). Algorithm 3 gives the outline of the search procedure for the aperture-based representation. The procedure starts by obtaining the minimum delta allowed to modify the intensity of an aperture (line 1). Each iteration, a random aperture of the selected angle is targeted in order to attempt improving the objective function of the treatment plan by either modifying its aperture shape or its intensity. When it is possible to perform both operations, the modification of the aperture intensity is selected with probability P_INT (line 8). As soon an improving solution is obtained, the algorithm returns it to re-start the process with a different beamlet. The aperture shapes are modified in line 11 and line 17. When the radiation must be reduced for the selected beamlet, if possible, we evaluate to close the beamlet by moving the right and the left leaf. Note that this implies that the procedure evaluates two modifications in some cases.

```
1  search (angle, beamlet, S); out: S'  δ ← getMinDelta(angle, beamlet, S);
2  S' ← S;
3  for a in randomOrder(angle.A) do
4      open ← isOpen(angle, beamlet, a);
5      if δ < 0 then
6          if !open then next;
7          if S.I[a][beamlet] + δ < 0 then  δ = −S.I[a][beamlet];
8          if rand() < P_INT then
9              |  S' ← modifyIntensity(angle, S, a, δ);
10         else
11             |  S' ← modifyAperture(angle, beamlet, S, a, δ);
12     else
13         if S.I[a][beamlet] + δ > MAX_P then  δ = MAX_P − S.I[a][beamlet] ;
14         if  open then
15             |  S' ← modifyIntensity(angle, S, a, δ);
16         else
17             |  S' ← modifyAperture(angle, beamlet, S, a, δ);
18     if z(x_{S'}) < z(x_S) then  return (S');
19     ;
20     S' = S;
21 return S'
```

Algorithm 3. Search procedure for the aperture-based representation.

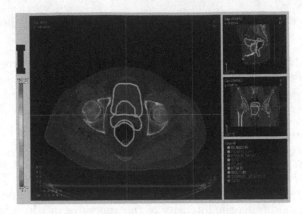

Fig. 5. Prostate case considered in this study. Two OARs (bladder and rectum) are considered. (Color figure online)

Perturbation: The perturbation procedure implements an operator that changes PSIZE times the initial solution S by randomly choosing, with probability 0.5, to modify the intensity of a randomly selected aperture a or modifying the aperture shape in a by opening or closing a randomly selected beamlet.

3 Experiments

A clinical prostate case obtained from the CERR package [7] has been considered in this study. Figure 5 shows a transversal view of this case. Boundaries of the target volume (tumor + margin in green), rectum (in light blue) and bladder (in yellow) are highlighted as the regions of interest in this study. We label the rectum and the bladder as *organs at risk* (OARs). The total number of voxels is about 56,000. The number of decision variables (beamlets) depends on the BAC and ranges between 320 and 380. The number of beam angles N considered in a BAC is equal to 5. The dose deposition matrix A^r is given for each BAC. We consider 72 beam angles, all of which are on the same plane. Just as in [4,6], we consider a set of 14 equidistant BACs to make our experiments.

We compare the algorithms described before to the sequential approach commonly used in clinical practice. That is, we first optimise the intensities (i.e. to solve the FMO problem) and then we find the set of deliverable apertures corresponding to the optimal intensities (i.e. to solve the MLC sequencing problem). While to solve the FMO problem we use the IPOPT solver [14], to solve the MLC sequencing problem we use the algorithm proposed in [1]. This algorithm finds a set of apertures delivering the optimal intensities and minimizing the beam-on time.

3.1 Parameter Configuration

In order to properly compare the proposed algorithms, we configure the parameter values of the intensity-based and aperture-based approaches using irace [12]. Each DAO algorithm run is allowed 60 seconds of execution time. The budget provided to irace for searching the parameter space of the algorithms was 1 000 evaluations and the configuration process was performed over 7 equidistant IMRT instances using default irace settings. The parameter settings obtained by irace, and used in the following experiments, are given in Table 1.

Table 1. Parameters settings obtained by irace (With the exception of N_RESTART which was set manually to 100 iterations) for the aperture-based algorithm (ABLS) and the intensity-based algorithm (IBLS).

Parameter	ABLS	IBLS	Parameter	ABLS	IBLS
BSIZE	7	45	P_INT	0.01	–
VSIZE	34	92	MAX_DELTA	–	15
PSIZE	6	6	α	–	0.99
Initial setup	open-min	open-min	MAX_RATIO	–	5
N_RESTART*	100	100			

3.2 Initialisation Method

The proposed algorithms have an intensifying nature due to the beamlet selection heuristic and the acceptance of only improving solutions. The initialisation setup of the aperture shapes and the intensities can influence the performance of the algorithms given that they provide the initial point from which solutions are repaired and improved. In the following experiments we evaluate the effect of the initialisation strategy on the performance of the proposed algorithms. For the aperture shapes we define 3 types of initial setup: (open, closed, random). The open setup initialises all the beamlets in the apertures open, closed does the opposite by initialising all beamlets as closed, and random initialises the aperture shapes randomly. The intensities can be initialised to the minimum intensity allowed (min), the maximum intensity (max), or a random (rand) value between the permitted minimum and maximum intensity. Table 2 gives the mean results obtained using the different initialisation approaches. In both algorithms the best mean results are obtained by the open-min strategy and the worst results are obtained by open-max. We note that the results of the closed-min strategy are not significantly different from the results of open-min for the intensity-based algorithm (p-value = 0.54). This hints that the selection of the most promising beamlet heuristic and the proposed search procedures benefit of a low initial intensity that will be iteratively increased during the search. Oppositely, repairing a treatment plan initialized with high intensities appears to be a more difficult scenario to tackle for these techniques.

Table 2. Mean value of $z(x)$ obtained by 30 repetitions of the algorithms using different initialisation strategies across the set of 14 BACs. The best mean results are shown in bold and results that are not significantly different to the bests mean results are shown in cursive (Wilcoxon signed-rank test with Bonferroni correction, $\alpha = 0.05$).

	open-min	open-max	closed-min	rand
IBLS	**55.2**	85.1	*55.6*	56.2
ABLS	**70.3**	137	98.8	91.0

3.3   Comparison of Performance

Table 3 shows a comparison between the different proposed strategies and the IPOPT solver which solves the FMO problem to optimality. The first column shows the BAC for which the intensities and deliverable apertures are computed. Columns 2–7 show the mean cost, the standard deviation (SD) and the mean beam-on time (BOT) found by the proposed intensity-based (IBLS) and aperture-based (ABLS) strategies considering a maximum of 5 apertures per beam angle, that is, a total of 25 deliverable aperture shapes per BAC. Each heuristic algorithm was run 30 times with a budget of 40 seconds on each instance. Columns 8–10 show the optimal cost for the FMO problem found by the IPOPT solver, the minimal beam-on time of the optimal solution found and a *lower bound* for the number of the aperture shapes required. The latter two values were reported by the MLC sequencing algorithm proposed in [1].

The results obtained by the proposed techniques are worst, in terms of the evaluation function, when compared to the optimal solution of the FMO problem. Nevertheless, the average beam-on time and number of aperture shapes required by this optimal solution are considerable larger. Large number of apertures and large beam-on times result in long treatment times which are undesirable from a practical perspective[2]. In particular, the significantly better beam-on time reported by IBLS is mainly due to the conditions imposed by the matrix representation of the intensities. For each angle, this matrix can be easily mapped to a small set of apertures with a total beam-on time equal to the maximum beamlet intensity in the matrix.

The solutions obtained by the intensity-based algorithm are competitive w.r.t the rounded solutions as they further reduce the beam-on time and number of aperture shapes increasing the applicability of the treatment plan and producing a smaller worsening of the evaluation function compared to the aperture-based technique. An important aspect of the desirability of treatment plans is the radiation dose distribution in the organs and tumours.

Figure 6 shows an example of dose distributions for an instance and the three strategies using their best configurations. Each curve represents an organ (the

---

[2] Even disregarding the impact on the clinical schedule, long treatment times are uncomfortable for the patients and carry an increased risk of intra-fraction motion, which may compromise plan quality [9].

Table 3. Mean results ($z(x)$), standard deviation (SD) and beam-on time (BOT) obtained by the intensity-based and aperture-based algorithms (IBLS and ABLS resp.), the FMO problem optimal solution evaluation (IPOPT, $z(x^*)$), the corresponding beam-on time and a lower bound for the required number of apertures (#ap).

Instances	IBLS			ABLS			IPOPT		
	$z(x)$	SD	BOT	$z(x)$	SD	BOT	$z(x^*)$	BOT	#ap
0-70-140-210-280	55.2	2.16	85	77.6	7.46	114	41.3	204	40
5-75-145-215-285	55.8	2.78	85	73.8	5.28	148	41.6	188	39
10-80-150-220-290	55.7	3.84	88	73.7	7.37	148	41.8	216	39
15-85-155-225-295	54.9	2.96	87	69.1	6.73	143	41.7	220	39
20-90-160-230-300	55.5	2.98	87	68.6	5.4	111	41.8	194	37
25-95-165-235-305	54.4	2.51	85	67.4	6.24	116	42.3	242	38
30-100-170-240-310	54.2	2.71	84	64.6	3.67	120	42.4	218	40
35-105-175-245-315	55.3	2.22	84	68.3	5.69	125	42.0	206	38
40-110-180-250-320	55.6	3.24	87	72.4	5.61	136	42.3	222	39
45-115-185-255-325	56.2	2.76	86	70.7	5.32	156	43.0	208	41
50-120-190-260-330	55.6	3.26	84	72.8	6.6	131	42.9	222	41
55-125-195-265-335	55.3	2.74	85	69.7	5.38	130	42.4	210	40
60-130-200-270-340	54.2	2.44	83	68.1	5.84	138	42.7	238	39
65-135-205-275-345	54.8	2.6	86	67.3	4.99	119	41.4	236	39
Average	**55.2**	**2.8**	**86**	**70.3**	**5.83**	**131**	**42.1**	**216**	**39.2**

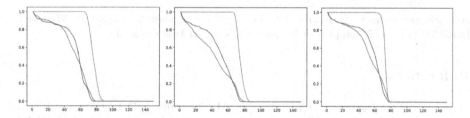

Fig. 6. Dose distributions of the instance 0-70-140-210-280 obtained by using IPOPT (left) the intensity-based strategy (middle) and the aperture-based one (right). (Color figure online)

yellow one corresponds to the tumor). Points (x, y) in each curve indicate that, related to the organ (or tumor), the $y\%$ of its voxels receive a radiation dose of at most x. Thus, it is desirable that the organ curves quickly decline to $y = 0$, while the tumor curve should remain high until the prescribed dose is reached.

Note that the dose distribution curves of the intensity-based strategy are quite similar to the ones reached by IPOPT. While, the tumor curve is slightly to the left, the organs clearly receive smaller radiation doses. On the other hand, in the aperture-based strategy the tumor reaches a dose closer to the prescribed dose but, as a result, the organs are affected to a greater extent by the radiation. Finally, highlight that, as our methods are stochastic, they may be able to offer different alternative treatments depending on the random number sequence.

4   Conclusions

In this paper we have introduced two novel heuristic algorithms to solve the DAO problem in radiation therapy for cancer treatment. Proposed heuristics are able to find a set of deliverable aperture shapes and intensities for each beam angle of a given BAC within a clinically acceptable time. Further, despite the heuristic algorithms were allowed to use only 5 apertures shapes per beam angle, they were able to find very competitive treatment plans.

We compare the heuristic algorithms to the traditional two-step approach where the optimisation of the intensities is performed first and then the sequencing problem is solved given an optimal intensity map. Results show that delivering the optimal plan would require more than two times the number of deliverable aperture shapes than our heuristic methods. These promising preliminary results encourage us to keep working on this research field.

As future work we want to include other criteria within our proposed framework, such as clinical markers. The collaboration of both techniques to solve DAO and the integration of the FMO problem model to use the optimal solution to guide this search is also part of our future work. Moreover, we also want to integrate to the proposed framework the beam angle optimisation problem, so we can (approximately) solve the whole IMRT problem in an integrated way.

Acknowledgement. Guillermo Cabrera-Guerrero wishes to thank FONDECYT/ INICIACION/11170456 project for partially support this research.

References

1. Baatar, D., Hamacher, H.W., Ehrgott, M., Woeginger, G.J.: Decomposition of integer matrices and multileaf collimator sequencing. Discrete Appl. Math. **152**(1–3), 6–34 (2005)
2. Broderick, M., Leech, M., Coffey, M.: Direct aperture optimization as a means of reducing the complexity of intensity modulated radiation therapy plans. Radiat. Oncol. **4**(1), 8 (2009)
3. Cabrera, G.G., Ehrgott, M., Mason, A., Raith, A.: A matheuristic approach to solve the multiobjective beam angle optimization problem in intensity-modulated radiation therapy. Int. Trans. Oper. Res. **25**(1), 243–268 (2018)
4. Cabrera-Guerrero, G., Lagos, C., Cabrera, E., Johnson, F., Rubio, J.M., Paredes, F.: Comparing local search algorithms for the beam angles selection in radiotherapy. IEEE Access **6**, 23701–23710 (2018)
5. Cabrera-Guerrero, G., Mason, A.J., Raith, A., Ehrgott, M.: Pareto local search algorithms for the multi-objective beam angle optimisation problem. J. Heuristics **24**(2), 205–238 (2018)

6. Cabrera-Guerrero, G., Rodriguez, N., Lagos, C., Cabrera, E., Johnson, F.: Local search algorithms for the beam angles selection problem in radiotherapy. Math. Probl. Eng. **2018**(1), 1–9 (2018)
7. Deasy, J., Blanco, A., Clark, V.: CERR: a computational environment for radiotherapy research. Med. Phys. **30**(5), 979–985 (2003)
8. Descovich, M., Fowble, B., Bevan, A., Schechter, N., Park, C., Xia, P.: Comparison between hybrid direct aperture optimized intensity-modulated radiotherapy and forward planning intensity-modulated radiotherapy for whole breast irradiation. Int. J. Radiat. Oncol. Biol. Phys. **76**, 91–99 (2009)
9. Dzierma, Y., Nuesken, F.G., Fleckenstein, J., Melchior, P., Licht, N.P., Rübe, C.: Comparative planning of flattening-filter-free and flat beam imrt for hypopharynx cancer as a function of beam and segment number. PloS One **9**(4), e94371 (2014)
10. Ehrgott, M., Güler, C., Hamacher, H.W., Shao, L.: Mathematical optimization in intensity modulated radiation therapy. Ann. Oper. Res. **175**(1), 309–365 (2009)
11. Jones, S., Williams, M.: Clinical evaluation of direct aperture optimization when applied to head-and-neck IMRT. Med. Dosim. **33**(1), 86–92 (2008)
12. López-Ibáñez, M., Dubois-Lacoste, J., Pérez Cáceres, L., Stützle, T., Birattari, M.: The irace package: iterated racing for automatic algorithm configuration. Oper. Res. Perspect. **3**, 43–58 (2016)
13. Shepard, D.M., Ferris, M.C., Olivera, G.H., Mackie, T.R.: Optimizing the delivery of radiation therapy to cancer patients. SIAM Rev. **41**(1), 721–744 (1999)
14. Wächter, A., Biegler, L.: On the implementation of a primal-dual interior point filter line search algorithm for large-scale nonlinear programming. Math. Program. **106**(1), 25–57 (2006)
15. Zeng, X., Gao, H., Wei, X.: Rapid direct aperture optimization via dose influence matrix based piecewise aperture dose model. PLOS One **13**(5), 1–11 (2018)

Hybridization of Stochastic Tunneling with (Quasi)-Infinite Time-Horizon Tabu Search

Kay Hamacher[(✉)]

Department of Computer Science, Department of Physics and Department of Biology,
Technical University Darmstadt, Schnittspahnstr. 10, 64287 Darmstadt, Germany
hamacher@bio.tu-darmstadt.de
http://www.kay-hamacher.de

Abstract. Stochastic Tunneling (STUN) is an optimization heuristic whose basic mechanism is based on reducing barriers for its search process between local optima via a non-linear transformation. Here, we hybridize STUN with the idea of Tabu Search (TS), namely, the avoidance of revisiting previously assessed solutions. This prevents STUN from inefficiently scan areas of the search space whose objective function values have already been "transformed away". We introduce the novel idea of using a probabilistic data structure (Bloom filters) to store a (quasi-)infinite tabu history. Empirical results for a combinatorial optimization problem show superior performance. An analysis of the tabu list statistics shows the importance of this hybridization idea.

1 Introduction

Real-world optimization problems often show multiple minima, e.g., several (local) optima. This occurs, for example, in the structure prediction of biomolecules where local minima correspond to partially folded proteins, or – more general – in any non-convex optimization scenario.

The number of those local minima typically increases with the dimensionality of the search space. Heuristics are a powerful tool to obtain approximate solutions in practice. Such approximations are at the same time better than only greedily (and thus locally) optimized solutions which are obtained by stubbornly following a (generalized) gradient and follow greedily only the path of strict improvements, eventually failing to overcome barriers between such (local) minima.

While Wolpert's "no free lunch" theorem [14] states that any optimization algorithm will not perform better – averaged over *all* optimization problems – than any other, for specific subset of optimization problems there are distinct performance improvements possible. This insight opens the route to *hybridization* techniques in which components of heuristic optimization procedures are synergistically combined. The possibilities of the hybridization idea are impressing due to the combinatorics of potential pairings of such components.

© Springer Nature Switzerland AG 2019
M. J. Blesa Aguilera et al. (Eds.): HM 2019, LNCS 11299, pp. 124–135, 2019.
https://doi.org/10.1007/978-3-030-05983-5_9

In this contribution we discuss the hybridization of a randomized search procedure (Stochastic Tunneling [6,7,13]) with the power of Tabu Search [3–5], which tries to avoid reassessing previously visited parts of the search space.

We will first review the details of both procedures and then propose our novel hybridization idea that overcomes shortcomings of both approaches applied individually. We will use the terms energy, objective function value, as well as objective function and potential energy surface interchangeably.

1.1 Stochastic Tunneling

Stochastic Tunneling (STUN) [6,7,13]) is a Monte-Carlo based sampling procedure in the spirit of Simulated Annealing (SA) [9]. While SA features a time-dependent temperature, STUN can be shown to be driven by an energy dependent (and thus position-dependent) temperature in certain cases [6].

In contrast to SA, STUN samples a transformed function f_{STUN} of the form

$$f_{STUN} = 1 - \exp\left(-\gamma \cdot (f(x) - f_0)\right) \tag{1}$$

with a transformation strength γ and the Metropolis-Monte-Carlo acceptance criterion based on a constant temperature β. Here, $f_0 := f(x_0)$ is the best known function value so far at the best guess of its location x_0. Note, that the transformation is *ad hoc* and does not imply the evaluation of $f(x)$ at several other locations (such as a Fourier-transform would require). STUN evaluates the objective function the same number of times as SA given finite iteration numbers i_{max} the procedure is allowed to run. In Fig. 1 we illustrate the effect of STUN.

The idea of STUN is to overcome barriers between minima more easily. This is possible as barriers are reduced whenever they are "higher" than the best known f_0 so far. The reliance on f_0 implies, that STUN is not Markovian any more, but eventually contains a *history of function values*.

The problem that plagues STUN after a considerable large number of steps is directly caused by the barrier reduction itself: when there are only a few better minima to be found, then the transformed function f_{STUN} becomes rather flat. Eventually, the STUN process diffuses on a flat surface with one remaining "hole" (the global minimum) – effectively playing golf.

In the extreme case when the structure and topology of the potential energy surface are disregarded, the processes is an unguided diffusion. Then, the visiting probability of any portion of the search space is uniformly likely and thus the density in search space uniform. Thus, in this worst case the process just guesses randomly.

It would thus be desirable to avoid not only function values that are uninteresting (which STUN does), but also areas of the search space previously visited. Then, the process is not a random walk anymore and guided by its *history in search space*.

Our idea, therefore, is to hybridize STUN to be dependent on the histories of both, the function values and the areas of search space visited.

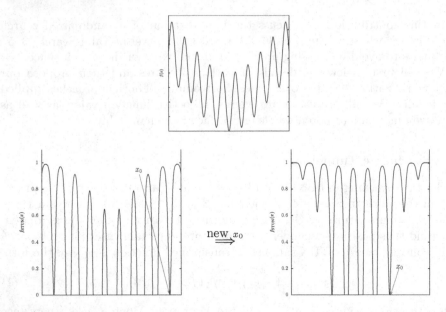

Fig. 1. The principle of the STUN transformation. Top: original objective function $f(x)$. Bottom, left: STUN transformation $f_{\text{STUN}}(x)$ of $f(x)$ for the indicated x_0 (red arrows); bottom, right: transformation after reaching an improved x_0. Note, how barriers are reduced upon a better estimate of x_0 and thus f_0. (Color figure online)

1.2   Tabu Search

From its inception on, Tabu Search (TS) [3–5], has gained tremendous traction in the heuristics community and is the procedure of choice to search with a history of areas visited during the course of an optimization run.

TS is also an iterative procedure that visits points in the search space based on a neighborhood relation (as SA and STUN). It, however, includes also a list of previously visited locations of the search space. Now, the number of visited locations needs to be limited due to memory and time/search constraints. This limit is called tabu tenure. Upon reaching this limit the history is either completely deleted or the "oldest" memory is progressively deleted.

The tabu region idea would be beneficial for STUN to avoid the diffusive behavior as argued above. However, a list of previously visited solutions of which entries are periodically revoked as the tabu tenure is reached might not address the diffusive behavior to its full extent. In theory, an infinite history would be the solution. However, this is unattainable in practice – again due to resource constraints.

2   (Quasi-)Infinite History via Bloom Filters

To implement a trade-off between a (sufficiently large) tabu history and the constraints of resources and look-up complexity, we propose a Tabu-STUN-hybrid

---

Algorithm 1. Storage procedure for the Bloom filter BLOOM for a configuration s in search space.

Require: k many distinct hash function $h_i(s)$ for members s of the search space $\mathcal{S}$
  function BLOOM.STORAGE(s)
    $v := (h_1(s), \ldots h_k(s))$
    for $1 \leq i \leq k$ **do**
      BLOOM[v_i] $:= 1$
  end function

Algorithm 2. Retrieval procedure for BLOOM; returns a Boolean to indicate whether s was seen before.

Require: k many distinct hash function $h_i(s)$ for members s of the search space $\mathcal{S}$
  function BLOOM.RETRIEVAL(s)
    $v := (h_1(s), \ldots h_k(s))$
    for $1 \leq i \leq k$ **do**
      if BLOOM[v_i] $== 0$ **then return** False
    return True
  end function

---

that relies upon a probabilistic data structure to store visited points in search space.

We propose to use Bloom filters [2] to represent the history of an STUN run. Bloom filters are of fixed size and store information of previously stored information. We use an array BLOOM of m bits to store information upon visited locations s. At start, each cell of BLOOM is initialized to zero. Each location to be entered into BLOOM is first mapped via k many hash functions to a vector-based fingerprint

$$v := (h_1(s), \ldots, h_k(s))$$

The hash functions h_i need not to be cryptographically secure, but must be distinct such that $h_i(s) \neq h_j(s)$ for $i \neq j$ almost always. For the search space $\mathcal{S}$ we have in addition $\forall i \in [1, \ldots, k], s \in \mathcal{S} : h_i(s) \in [1, \ldots m]$. We then can use the individual values $h_i(s)$ as addresses of BLOOM. We enter a new visited location by setting every bit in BLOOM addressed by the fingerprint vector v. We ask whether we have seen a s by mapping again to addresses $h_i(s)$ and taking a logical **and** over the binary values under the k many addresses thus computed.

Note, that BLOOM's recall is exact: any input ever encountered before will be reported as such; however, to avoid infinite storage requirements, a Bloom filter shows a non-vanishing error rate in reporting false positives. Thus, a Bloom filter will assign a "visited" label to some configurations in search space that have not been assessed before. In Algorithms 1 and 2 we show the procedures of storage and retrieval in pseudo-code.

In [12] the rule $k := \frac{m}{d} \log_e 2$ is derived as an optimal value for k based on a minimal false positive rate p. The Bloom filter has given size m and data $\{s_1, \ldots, s_d\}$ is to be inserted. Note, that this relation implies a linear increase in storage size m with respect to data size d. While the scaling in requirements is the same, the amount of real memory needed, however, could be orders of magnitude different: m is the number of *bits* stored in the Bloom filter, while a datum s might be a complex data structure with large memory footprint, e.g., in our test system a datum s needs at least 400 bits for its representation (see Sect. 3.1 below). Thus, a Bloom filter can help to reduce memory requirements tremendously – eventually, enabling a "full-take" of the search history.

2.1   Proposed Hybrid Algorithm

In Algorithm 3 we show pseudo-code for our proposed Bloom-augmented STUN variant. We create new candidate solutions t for the Monte-Carlo process until we find a never visited one (assured by the condition `BLOOM.Retrieval(t)`).

Due to potential false positives and/or a "trapping" in a corner of the search space, we must, however, also allow the process to escape such a trap. We achieve this by the additional condition that no more than $2 \cdot |s|$ many trials are tested for.

Eventually, this limit was never reached in our application (see below, Sect. 4). Therefore, this condition remains a hypothetical assurance for the process to not get stuck in an infinite loop that in practice is never realized.

The history of solutions visited and stored in the Bloom filter is (quasi-) infinite when we use the number of maximal iterations i_{max} in STUN as the size of the data to be put into the Bloom filter $d = i_{max}$. Here, the Bloom-filter can be allocated beforehand as we know i_{max} in advance.

2.2   Potential Shortcomings

Avoiding diffusive behavior of STUN via Bloom filters might lead to ignorance about some configurations (false positives) and – in the worst case – prevent our new STUN variant to ever reach the global minimum. Thus, the Bloom filter-based step generation is a trade-off whose advantages and disadvantages need to be assessed for a particular application. To this end we have chosen a combinatorial optimization problem described in the next section.

3   Experimental Setup

3.1   Test Case

We need a test instance whose objective function can be evaluated very fast, but which shows all the characteristics of difficult optimization problems (multi-minima, exponential scaling of those with the dimension of the test instance, large to arbitrary barriers between local minima). Ising spin-glasses [1] with Gaussian distributed interaction couplings fulfill all these specifications, while at the same time exact solutions can be computed by an external sources [10]. The objective function reads

$$\min E(s) = \sum_{<i,j>} J_{ij} s_i s_j \qquad \forall i \in \{1...N\} \; : \; s_i \in [-1/2; 1/2] \qquad (2)$$

Algorithm 3. Proposed STUN-variant with (quasi-)infinite tabu tenure; using the STUN transformation function $f(x,y) := 1 - \exp\left(-\gamma \cdot (x - y)\right)$. Additions to the original STUN algorithm of Refs. [7, 13] are marked in blue.

Require: k many distinct hash function $h_i(s)$ for members s of the search space $\mathcal{S}$
Require: neighborhood relation $\mathcal{N}(s)$ for any s of the search space $\mathcal{S}$
Require: rnd() uniform random number generator $\in [0; 1]$
Require: s_0 some starting point
Require: $E(s)$ objective function
Require: β, γ as constant hyperparameters
Require: initialized Bloom-filter object BLOOM
  $E_0 := E(s_0)$, $s := s_0$, $E := E_0$
  for $1 \le i \le i_{\max}$ **do**
    $w := 0$
    repeat $\triangleright$ find previously non-visited t
      $t :=$ draw from $\mathcal{N}(s)$ uniformly
      $w := w + 1$
    until $w > 2 \cdot |s|$ or not BLOOM.Retrieval(t)
    insert w into histogram $h(w)$ $\triangleright$ optional: for analysis in Fig. 4
    if $\exp\left(-\beta \cdot [f_{\text{STUN}}(E(t), E_0) - f_{\text{STUN}}(E, E_0)]\right) <$ rnd() **then** $\triangleright$ accept move
      $s := t$
      $E := E(t)$
      BLOOM.Storage(t) $\triangleright$ store a newly visited t
      if $E < E_0$ **then** $\triangleright$ better solution found
        $E_0 := E$
        $s_0 := t$
  return (s_0, E_0) $\triangleright$ best spin configuration and its energy

---

The sum over $<i, j>$ of Eq. 2 is restricted to nearest neighbors on a 2D grid of N spins and thus a regular lattice of side length $\sqrt{N}$. The J_{ij} are drawn from a Gaussian distribution. The vector $s^* := (s_1^*, s_2^*, \ldots, s_N^*)$ contains the optimal solution. We will deal with a 20×20 grid of spins, thus with $N = 400$. This problems has a large sample space[1] of size 2^{399}.

3.2   Experimental Details

The algorithm was implemented in C++, using g++ and the hash-function of the Boost library based on existing Ising simulation code [6, 8]. Production runs were done under the Linux OS. The k many different hash function h_i were derived by k copies of Boost's hash function and seeding each with a distinct random number.

The computational burden in real-time units (like seconds etc.) is largely influenced by hardware details and software versions. To circumvent this problem we only refer to the number of calls to the objective function $E(s)$ as a proxy for the time spent in the overall computation.

---

[1] For general, random J_{ij} there exist one symmetry between up/down-spin states that eventually degenerates into two global solution dividing the search space in one half.

We show the *average* performance of 250 independent runs of the Ising ground state problem: obtained for 50 independently created Ising spin systems (50 times drawn N values J_{ij}) for which we apply STUN and STUN-BLOOM five times independently for each instance.

3.3 Hyperparameters

The inverse temperature β and the transformation strength γ are the hyperparameters of STUN. Furthermore, the number of distinct hash functions k used in the Bloom filter can be regarded as another one. However, analytic arguments [12] determine the optimal value for $k := \frac{m}{d} \log 2$ given the Bloom filter size m and the number of data d. Now, d could be chosen larger than necessary. Here, we decided to set $d = i_{\max}$ as the number of iterations. Thus, we demand the Bloom filter to contain $i_{\max}$ configurations at most while not wasting memory. Our setup is thus optimize for low(er) memory footprint while being able to distinguish most if not all visited configurations (but not more). From this point of view, the history taken into account is effectively (quasi-)infinite for a prescribed number of iterations.

For $i_{\max} = 2 \cdot 10^8$ and $p = 10^{-8}$ we obtain $m = 7,668,046,702$ which amounts to a memory demand 914.1 MiB = 958.5 MB using the optimal $k = 27$.

We performed a scan for suitable hyperparameters β and γ taken from the grid $(\beta, \gamma) \in [0.1; 1; 5; 10] \times [0.1, 1, 10]$ for $i_{\max} = 2 \cdot 10^8$. We found $\beta_{\text{opt}} = 5$ and $\gamma_{\text{opt}} = 10$ to perform best with respect to the average relative error of the objective function value $\epsilon_{\text{rel}}(n_{\text{iter}})$ which we define as

$$\epsilon_{\text{rel}}(n_{\text{iter}}) := \frac{E^{(\alpha)}(n_{\text{iter}}) - E_o^{(\alpha)}}{E_o^{(\alpha)}}$$

where $E_o^{(\alpha)}$ is known global optimum value obtained from the spin-glass-server [10]. α is the id of the respective test-instance; n_{iter} the iteration number – corresponding to the number of calls of the objective function.

4   Results

Note, that false positives in the Bloom filter do not decrease the number of visited configurations, but rather "push" the search process further away than necessary. By this construction we ensure that exactly $i_{\max}$ distinct spin configurations are visited. We never reached the upper limit of too many visited configurations as tested for in Algorithm 3.

In Fig. 2 we show a direct comparison of the classical STUN procedure with the herein proposed variant based on Bloom filters. We used raw data for all hyperparameters and iteration numbers. As is evident, the new variant almost always performs better than the classical STUN.

For larger iteration numbers, the superiority of the new algorithm is even more pronounced as can be seen in Fig. 3 where we show the averaged relative

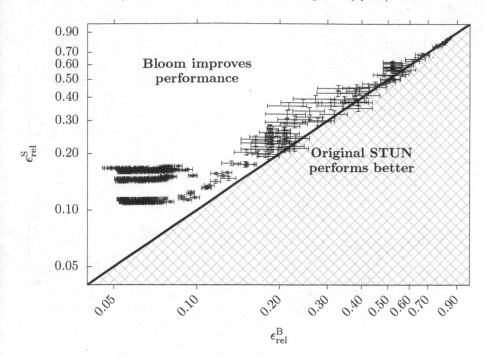

Fig. 2. Averaged relative errors ϵ_{rel} compared for identical hyperparameters (β, γ) and various iteration numbers i between STUN (S) and BLOOM-STUN (B). Error bars represent standard deviations over 250 independent runs and replicas as described in Sect. 3.3 over which also the averages where taken. We indicate by the filled triangle situations for which STUN performed better as STUN-BLOOM (negligible).

error as a function of the number of calls of the objective function for the best hyperparameters of each variant.

The drop at a distinct number of iterations in Fig. 3 was observed previously [7,8]. It is caused by reaching a local minimum from the arbitrary starting configuration. Immediate gains are thus easy during the first $\sim 10^5 - 10^6$ iterations. The ultimate challenge, however, is the search dynamics during later times ($10^6 - 10^8$ iterations). Clearly, Fig. 3 shows that BLOOM-STUN does improve upon STUN. While STUN for various parameters seem to get stuck, the new Bloom filter-based variant has not reached any saturation.

To illustrate the importance of the `BLOOM.Retrieval` condition in Algorithm 3 we went further and analyzed how many candidate configurations t were actually tested for membership in `BLOOM`.

The answer is somewhat surprising: on average $\bar{w} \sim 3.8$ neighbors needed to be tested to find a candidate t never visited before. In Fig. 4 we present the histogram $h(w)$ of Algorithm 3.

From these observation on $h(w)$ we can immediately conclude that the initial motivation to incorporate a (quasi-)infinite Tabu Search-like avoidance of previously visited solutions was justified: without this mechanism the classic

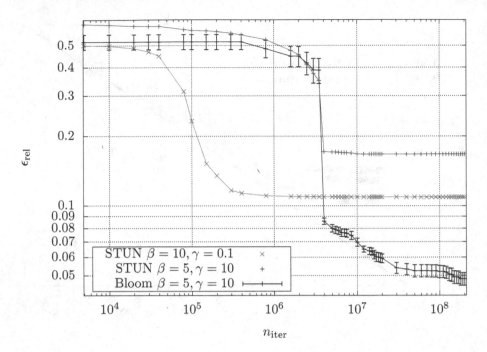

Fig. 3. Averaged relative errors ϵ_{rel} as a function of computational time (expressed as the number of calls to the objective function). Error bars for BLOOM-STUN represent standard deviations over 250 independent runs and replicas as described in Sect. 3.3. Error bars for STUN are of the same order, but were omitted for visual clarity. We directly compare the performance of STUN and BLOOM-STUN at BLOOM-STUN's best hyperparameters $(\beta, \gamma) = (5, 10)$ [black and purple], as well as for STUN's best hyperparameters $(\beta, \gamma) = (10, 0.1)$ [green]. (Color figure online)

STUN process would have revisited time and again already assessed areas of the search space. With the inclusion of a tabu mechanism we have forced the search space out of the seemingly diffusion-like behavior into one that searches more effectively.

This conclusion is furthermore strengthened by the different choices of best hyperparameters (as shown in Fig. 3): the tabu/Bloom based variant runs better at a larger temperature (smaller β) and stronger transformation (larger γ) than the classic STUN. The original Monte-Carlo process without a tabu list would be much more diffusive, thus more inclined to randomly guess than the Bloom/tabu list augmented one.

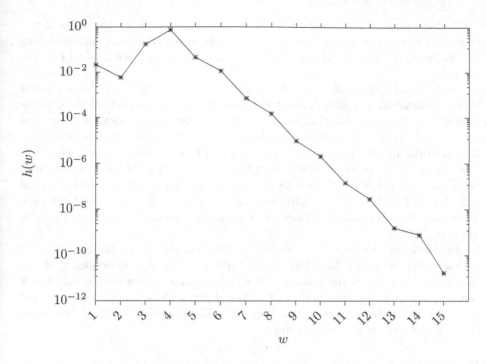

Fig. 4. Histogram $h(w)$ of the number of tested Ising configurations w in each STUN/Monte-Carlo step of Algorithm 3. Note, that the final step was always successful, that is a new configuration never encountered before. Thus, a history-unaware, Markovian process would show a delta peak at w. Here, we averaged over 300 independent runs.

5   Conclusions

Motivated by the observation that the heuristic optimization procedure Stochastic Tunneling (STUN) tends to diffusion-like (and thus inefficient) search behavior, we hybridized it with the idea of Tabu Search (TS) to avoid re-visiting areas of the search space again and again.

We used the probabilistic data structure of Bloom filters to maintain a (quasi-)infinite history of visited solutions. A Bloom filter is a trade-off between reducing the memory requirements to store the full search history and a (small and *a priori* setable) false positive rate.

This small modification of STUN rendered the search process exponentially more effective in regard to the relative error for a combinatorial optimization problem (Ising spin glass ground states).

In theory, the usage of a probabilistic structure with a false-positive rate could potentially lead to exclusion of the global minimum. In practice, however, where there myriads of local minima, this hypothetical outcome seems unimportant when contrasted with the beneficial performance shown in Fig. 2.

Furthermore, our *ex post* analysis of the importance of the tabu list – expressed in the histogram $h(w)$ of the number of rejected tabu configurations w – showed that the initial hypothesis on the improvable details of STUN was correct.

As a final note, we want to critically remark that the look-up via BLOOM.Retrieval might be more time-consuming than the actual recomputation of the objective function itself. Clearly, in our test system of Ising spin glasses the overall computational burden lies not in the objective function itself. In real-world applications of heuristics, such as protein structure prediction [11], the relation is, however, reversed: here the dominant part of the computations reside in the evaluation of the objective function and not in the rather trivial BLOOM.Retrieval procedure with some $k \approx \mathcal{O}(10)$ hash function evaluations. Therefore, in such realistic settings the additional overhead to maintain the tabu list is negligible.

In the future, we want to derive an analytic theory for the findings on $h(w)$ in Fig. 4. We speculate that the rather high $\bar{w} \sim 3.8$ is a consequence of the high-dimensionality of the Ising search space in combination with a only a few (namely two) choices per degree of freedom. Such a theory, however, is beyond the scope of the hybridization question and will be pursued in the future from the point of view of stochastic processes.

References

1. Binder, K., Young, A.: Spin glasses : experimental facts, theoretical concepts, and open questions. Rev. Mod. Phys. **58**(4), 801–976 (1986)
2. Bloom, B.H.: Space/time trade-offs in hash coding with allowable errors. Commun. ACM **13**(7), 422–426 (1970). http://doi.acm.org/10.1145/362686.362692
3. Glover, F.: Future paths for integer programming and links to artificial intelligence. Comput. Oper. Res. **13**(5), 533–549 (1986)
4. Glover, F.: Tabu search-part I. ORSA J. Comput. **1**(3), 190–206 (1989)
5. Glover, F.: Tabu search-part II. ORSA J. Comput. **2**(1), 4–32 (1990)
6. Hamacher, K.: Adaptation in stochastic tunneling global optimization of complex potential energy landscapes. Europhys. Lett. **74**(6), 944–950 (2006)
7. Hamacher, K., Wenzel, W.: The scaling behaviour of stochastic minimization algorithms in a perfect funnel landscape. Phys. Rev. E **59**(1), 938–941 (1999)
8. Hamacher, K.: A new hybrid metaheuristic – combining stochastic tunneling and energy landscape paving. In: Blesa, M.J., Blum, C., Festa, P., Roli, A., Sampels, M. (eds.) HM 2013. LNCS, vol. 7919, pp. 107–117. Springer, Heidelberg (2013). https://doi.org/10.1007/978-3-642-38516-2_9
9. Kirkpatrick, S., Gelatt, C., Vecchi, M.: Optimization by simulated annealing. Science **220**, 671–680 (1983)
10. Simone, C., Diehl, M., Jünger, M., Mutzel, P., Reinelt, G.: Exact ground states of ising spin glasses: new experimental results with a branch-and-cut algorithm. J. Stat. Phys. **80**, 487 (1995)
11. Strunk, T., et al.: Structural model of the gas vesicle protein GvpA and analysis of GvpA mutants in vivo. Mol. Microbiol. **81**(1), 56–68 (2011)
12. Tarkoma, S., Rothenberg, C.E., Lagerspetz, E.: Theory and practice of bloom filters for distributed systems. IEEE Commun. Surv. Tutorials **14**(1), 131–155 (2012)

13. Wenzel, W., Hamacher, K.: A Stochastic tunneling approach for global minimization. Phys. Rev. Lett. **82**(15), 3003–3007 (1999)
14. Wolpert, D.H., Macready, W.G.: No free lunch theorems for optimization. IEEE Trans. Evol. Comput. **1**(1), 67–82 (1997)

A Self-adaptive Differential Evolution with Fragment Insertion for the Protein Structure Prediction Problem

Renan S. Silva and Rafael Stubs Parpinelli[✉][iD]

Graduate Program in Applied Computing, Santa Catarina State University,
Joinville, SC, Brazil
renan.silva@edu.udesc.br, rafael.parpinelli@udesc.br

Abstract. This work presents four hybrid methods based on the Self-adaptive Differential Evolution algorithm with fragment insertion applied to the protein structure prediction problem. The protein representation is the backbone torsion angles with side chain centroid coordinates. The fragment insertion is made by the Monte Carlo algorithm. The hybrid methods were compared with recent and compatible methods from the literature, where two proposed approaches achieved competitive results. The results have shown that using parameter control and fragment insertion greatly improves the results of the prediction when compared to fragment-less methods or without parameter control. Furthermore, an extra analysis was conducted using GDT-TS and TM-Score metrics to better understand the results obtained.

Keywords: Structural biology · Bioinformatics
Evolutionary algorithms · Parameter control · Monte Carlo search

1 Introduction

Proteins are one of the four macromolecules essential for the life as we know it. They are responsible for metabolic, structural, hormonal, regulatory and other functions. The three-dimensional conformation of a protein (i.e, its shape, structure) has a direct connection to the protein function. Therefore, knowing the three-dimensional conformation can give insights on the roles that a protein has in an organism, on drug design and a better understanding of diseases [24]. Each protein can be uniquely identified by its amino acid sequence. The process of determining this sequence is called protein sequencing and it is a relatively cheap process. On the other hand, determining a protein structure depends on expensive methods as X-Ray crystallography or nuclear magnetic resonance, which are slow, error-prone and very expensive [6]. Hence, the gap between the number of sequenced and structured proteins is larger than 3 orders of magnitude [2].

With the goal of closing this gap, scientists have been working for decades to model methods to predict the protein structure from its sequence. This problem is referenced as Protein Structure Prediction Problem (PSPP) and it is

M. J. Blesa Aguilera et al. (Eds.): HM 2019, LNCS 11299, pp. 136–149, 2019.
https://doi.org/10.1007/978-3-030-05983-5_10

considered one of the main open problems in computer science and bioinformatics [5,11,12]. Several protein representation models have been proposed to handle the PSPP in different levels of details. The HP on-lattice model, which can be considered as the simplest one, is an $\mathcal{NP}$-Hard problem [1,9] being unfeasible to solve it for real instances using exact methods. One attempt at solving the HP model is presented in [15], however, they are far from achieving good results for real proteins. Other models used in the literature describe the protein as a set of backbone angles and torsions with the option of using the side chain information. These representations have a higher level of details at the cost of complexity, thus requiring the use of (meta)heuristics methods, such as bio-inspired algorithms [17]. Examples of such works can be seen in the use of Genetic Algorithms [3], Memetic Algorithms [7], Simulated Annealing [22], hybrid methods [29], and Differential Evolution [14]. The Differential Evolution (DE) has shown its potential over the years and it is currently one of the best performing continuous optimizers.

It is known that parameters play a big role in the performance of metaheuristics [18], and this is no different for the PSPP. Finding the optimal set of parameters is usually a long process and very resource consuming since it requires many executions of the optimization process. Furthermore, the different target protein may have different optimal parameters. Therefore, finding the parameters prior to making the final prediction is not ideal resource-wise. Another problem also arises from this: Using only one set of parameters during the whole process may impact negatively on the optimization process, that is related to the balance between exploration and exploitation. And, different steps of the process may require a different set of parameters to achieve a better performance. Works such as [14] shows that even hard coded strategies for different steps of the process can improve the final solution. It has been shown in several studies that the choice of parameters is critical for the performance of DE [10].

Some algorithms can be considered for parameters' auto adaptation using DE: JADE [28], CODE [25] and SaDE [20]. The one chosen to be used in this work was SaDE, for the reasons explained next. SaDE is capable of adapting its parameters over the course of the optimization, based on each operator in use. It can adapt the parameter over the run, allowing for the use of heterogeneous operators working together. The other reason for using SaDE is that it can control which operator will be used, based on its success rate. This allows for SaDE to use different operators over different stages of the optimization process.

This work investigates the use of a hybrid approach based on the Self-Adaptive Differential Evolution algorithm coupled with fragment insertion in four variants. The fragment insertion is made by the Monte Carlo algorithm [23]. The goal is to adapt the search to the problem complexity during the optimization process while including more knowledge from the problem domain. This is accomplished with the use of a fragment library assembly using Rosetta [16]. A fragment library is a set of protein pieces of which all are a sub-sequence of the target protein, in a way that it can be assembled (or at least approximated) with a set of fragments. Also, we provide an analysis of the proposed algorithms using several metrics, with the objective of making more information about their performance and allowing future developments to compare to our results obtained.

This work is structured as follows. Section 2 presents the PSPP, the computational model used and the target energy function. Section 3 presents the Self-adaptive Differential Evolution used in this work. Section 4 shows the hybrid fragment-based approaches proposed in this work. In Sect. 5 the experimentation setup used is shown. In Sect. 6 the results obtained are presented and analyzed. Section 7 contains the conclusions and future research directions.

2   The Protein Structure Prediction Problem (PSPP)

A given protein can be analyzed at different levels of detail: Primary, secondary, tertiary and quaternary. At the primary level, the protein structure is analyzed considering only the amino acid sequence, which can be used to uniquely describe the protein. The secondary structure considers recurrent patterns inside a local sequence of amino acids considering the dihedral angles, where the most common ones are the α-helix (helicoid shape), β-sheet (planar shapes) and coils (irregular shapes). The tertiary structure corresponds to the three-dimensional shape of the protein, also called the three-dimensional conformation or native conformation. The quaternary structure is formed by the composition of 2 or more proteins. Thus, the PSPP can be defined as finding the tertiary structure of a protein from its primary structure.

2.1   Protein Representation

When representing a protein computationally there are several options available, depending on the level of detail desired and what is being considered about the protein. Those models can then be divided into two major classes: On-lattice and off-lattice. The on-lattice models consider each amino acid as a point in space and the bonds between them are lines. These lines are restricted to a regular lattice in a two or three-dimensional space. It is possible to solve the on-lattice models exactly and to prove its optimality [15], however, their lack of details makes them far from real conformations. Off-lattice models, on the other hand, can represent proteins in a higher level of details, since its angles are not limited to a fixed increment. Its main models are: C_α Coordinates, all heavy atoms coordinates, backbone torsion angles with side chain centroid coordinates, backbone and side chain torsion angles, and all atoms coordinates. This research uses the backbone torsion angles with side chain centroid coordinates model, as illustrated in Fig. 1. Each peptide bond is composed of 3 angles, *phi*, *psi* and *omega*, that can rotate freely from $-180°$ to $180°$. Each angle represents a variable to be optimized.

2.2   Energy Landscapes

The process of predicting a protein tertiary structure based only on the primary structure is referenced in the literature as *ab initio*. This process can be

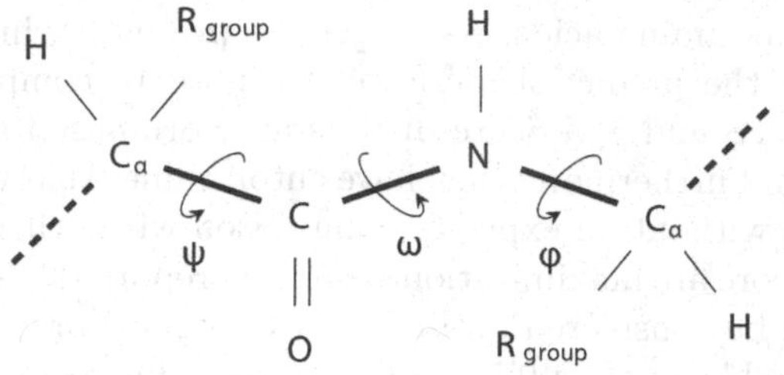

Fig. 1. Backbone protein representation

improved by utilizing more knowledge gathered from other sources of information than the primary structure. Examples of this are the prediction of the secondary structure, libraries of rotamers, information about the angle distribution, and libraries of fragments. When some of this information is utilized we have a *de novo* prediction. Either of those methodologies are guided by the optimization of a potential energy function, such as AMBER, CHARMM, or Rosetta [5]. These energy functions are computable approximations of the many physical and chemical interactions that happen naturally. They are extremely complex, requiring advanced methods to overcome the high dimensionality and multimodality of the problem. This work uses Rosetta's energy function, Rosetta's fragment library and secondary structure prediction with PSIPRED [4].

2.3 Prediction Assessment

To evaluate how good (or bad) a prediction is, it is necessary to have an assessment metric. The main assessment metric is based on the Root Mean Square Deviation (RMSD), as shown in Eq. (1). It consists of iterating two conformations A and B over all n α-carbons, squaring their differences in position, averaging it according to n and then returning the square root of this value. It is worth noting that the two conformations must be aligned.

$$RMSD_\alpha(A, B) = \sqrt[2]{\frac{\sum_{i=1}^{n} (A_i - B_i)^2}{n}} \tag{1}$$

Most of the works found in the literature use only the RMSD as an evaluation metric. However, there are many other metrics that can be employed for a better (or complimentary) assessment than the RMSD. Two other metrics are the GDT-TS [27] and the TM-Score [30], which are largely utilized in CASP [13] competition. Besides RMSD, both GDT-TS and TM-Score metrics will be used in this work.

Both GDT-TS and TM-Score have several advantages over RMSD. They can provide a normalized value and are less sensitive to local changes, while RMSD can be very sensitive to local variations. RMSD also scales quadratically, which can make values hard to be interpreted. Also, the RMSD is sensitive to the protein number of amino acids. As bigger the protein is, higher the RMSD can be. This makes the prediction performance hard to compare on different proteins. The GDT-TS and TM-Score metrics are normalized and nonsensitive to the protein length. Furthermore, they have cutoff values that can indicate how good a prediction is without an explicity comparison with other methods. Both GDT-TS and TM-Score are maximization scores. As reported in [26], a TM-Score of 0.5 or bigger can be considered an indicator of a good prediction. Similarly, for GDT-TS, a value bigger than 0.5 can be considered a good prediction, while a value less than 0.2 indicates the performance of a random search.

3 Self-adaptive Differential Evolution (SaDE)

The Differential Evolution (DE) algorithm is a metaheuristic proposed by Storn and Price in 1996 for continuous optimization problems [19]. It consists of a population of size NP of solution vectors of size D, where new individuals are generated by recombination of individuals of the previous generation. Each individual has a fitness value proportional to how good it is and this value is responsible for guiding the DE in the search space. A newer individual compete with the previous one and must have a better fitness in order to replace the competing individual.

There are several means of combining individuals to form newer ones. Table 1 presents the ones used in this work. The operators are named as $xx/yy/zz$ and mean the following. The term xx names what is used as a base vector. Can be *rand* meaning a random vector x_{r1}, *best* meaning the vector with the best fitness x_{best} or *curr* meaning the current vector x being iterated. The second term, yy, is the number of difference vectors used, and it is usually 1 or 2. The third term, zz, is the combination operator. Two main methods of combining vectors are present in the literature: the binary operator and the exponential operator. In short, the binary chooses for each dimension the source of the mutation, whereas the exponential operator copies a continuous sequence. Given the nature of the PSPP, copying continuous sequences of amino acids can lead to better results.

Table 1. DE mutation operators

Name	Formulation
best/1/exp	$w = x_{best} + F \times (x_{r1} - x_{r2})$
best/2/exp	$w = x_{best} + F \times (x_{r1} - x_{r2}) + F \times (x_{r3} - x_{r4})$
rand/1/exp	$w = x_{r1} + F \times (x_{r2} - x_{r3})$
rand/2/exp	$w = x_{r1} + F \times (x_{r2} - x_{r3}) + F \times (x_{r4} - x_{r5})$
curr-to-best/2/exp	$w = x + F \times (x_{best} - x_x) + F \times (x_{r2} - x_{r3})$
curr-to-rand/1/exp	$w = x + F \times (x_{r1} - x_{r2})$

DE has 3 main parameters: F, Cr and NP. F is a scalar that controls how much the base vector is perturbed compared to the differential vectors used. A smaller value leads to a local search (or exploitation), where a bigger value leads to a more global search (or exploration). The parameter Cr controls how much of the current vector is replaced by the new one. A small value leads to only a few variables being replaced, while a bigger value will make more variables being replaced. This can be used to speed up or hold down the convergence speed. Finally, NP simply control how many solution vectors there are in the population. A bigger value will mean that the search space is more explored inside the same generation and a smaller value leads to only a few examples of the search space. Assuming a fixed budget of function evaluations, a bigger value leads to fewer generations, which in turn gives the individuals fewer opportunities for evolution.

The SaDE algorithm [20] is responsible for controlling 3 aspects of DE. First, it modifies F using a normal distribution with mean 0.5 and standard deviation 0.3. This parameter is not adapted but suffers random variations favouring exploration and exploitation in all stages of the optimization process. The Cr parameter is controlled by the SaDE in which each operator available to SaDE has its own value of Cr. Hence, different operators may have different optimal parameters. The adaptation of Cr is based on a learning phase, which tries to find good values for it. After the learning phase, the median of the best values is used as Cr. Each time an operator is applied a new value is generated based on the median, which allows for the parameter to slightly change and adapt over time. If the newly generated value was able to lead to an improvement, then it is stored for later use. The third aspect under SaDE is the mutation operator. The original work of [20] uses 5 operators. In this work, it ranges from 4 to 10. Each time an operator is applied to a solution vector, its success (or failure) is recorded. Over time, the operators which had the bigger amount of improvements has a bigger chance of being used than the ones with a smaller one based on a proportional selection. With that, SaDE is able to use the operators that are most likely to have a positive effect while avoiding spending function evaluations on the bad ones. For more details on SaDE see [20,21].

4   SaDE with Fragment Insertion

This work explores four variations of the hybrid approach based on the SaDE algorithm with the use of fragment insertion. The fragment insertion is made by the Monte Carlo algorithm [23]. Since SaDE is able to control the DE parameters as well as select which operators to use, our hybrid approach consists in putting problem specific operators under SaDE control to achieve hybridization, bridging the gap between the optimizer and the problem.

Fragments are pieces of proteins that had its structure determined in a laboratory and which have the same subsequence as out protein. There are 4 fragment insertion operators: *3mer*, *3merSmooth*, *9mer* and *9merSmooth*. The *3mer* and *9mer* consist of a simple fragment insertion of size 3 or 9, respectively. On the other hand, *3merSmooth* and *9merSmooth* can add a fragment into a random position, and them, select a second fragment insertion in a way where the total backbone displacement is reduced. These two operators can achieve a greater acceptance rate (since it is less likely to make a bad move) and can explore the search space more locally. These operators use experimental data to assemble the fragment library, thus, they can give more problem domain information. For this, two new set of domain-specific operators are presented.

The first set of domain-specific operators used in the proposed hybrid approach consists in pure fragment insertion and will be referenced as the FRAG operator. As mentioned before, their insertion is made by the Monte Carlo algorithm that is applied for a predefined number of attempts. Hence, improving changes are always accepted and worsen changes are probabilistically accepted according to a predefined constant, also known as temperature. This constant is controlled by the SaDE algorithm. This criterion allows the operator to escape from local minima.

The second set of domain-specific operators is very similar to the FRAG operator, having all of its elements. While FRAG is the operator set with the biggest amount of information available, it also has no capacity to use the information stored in other solution vectors (individuals). That is, if there is information of good quality available at one solution vector, the others have no access to it. One way of dealing with that is to use the Replica Exchange Monte Carlo (REMC) [8], which can copy the solution vector of other solution vectors before starting applying the fragments. This permits reuse of good information between different solution vectors. This approach will be referenced as REMC operator. Each time the operator is called, it competes with the previous solution vector. That is, if the i-th solution vector had the REMC operator called, first, it will be compared to the $(i - 1)$-th solution vector with the Monte Carlo criterion. This wraps around, so the first solution vector competes with the last one. This means that the information is passed in a ring topology, thus, preventing premature convergence of the algorithm while at the same time sharing good solutions vectors information.

In light of this, the four hybrid methods explored in this work are SaDE with FRAG and DE operators; SaDE with REMC and DE operators; SaDE using only FRAG operators; SaDE using only REMC operators. The reasoning in exploring these four optimization models are as follows. SaDE is able to do parameter control during the optimization process and can select which operators are given the most improvement. Both FRAG and REMC operators are able to include problem-specific information into the optimization process, giving to DE the ability to better explore the potential energy landscape.

Finally, all four optimization models have some common elements. All solution vectors are codified in terms of backbone angles, three for each peptide, that can rotate freely from $-180°$ to $180°$. The initial population is generated by optimizing a simple objective function by using fragment insertion. This function is named *score0* (available in Rosetta) and consists only of repulsive van der Waals force, which is capable to detect conformations with colliding parts and are undesirable. At the end of the optimization process, the conformation with the best energy score is selected and repacked. Repacking consists of replacing the side chain centroids by its original structure. This requires a re-optimization of the protein backbone, which is conducted with gradient descent.

5 Experiments Setup

With the goal of verifying the performance of the proposed optimization models, 4 proteins were selected from PDB [2] based on their utilization in the literature. The protein properties are summarized in Table 2, where the first column holds the protein name, the second column shows the protein size (number of amino acids), the third column holds the number of backbone angles (which corresponds to the number of variables being optimized), and the fourth column shows the secondary structures found in the native conformation of the protein.

Table 2. Target proteins and their features

Name	Size	Backbone angles	Structure
1ZDD	35	105	2α
1CRN	46	138	$2\alpha, 2\beta$
1ENH	54	162	3α
1AIL	72	216	3α

The main objective of the optimization process is to minimize the score function, thus, this is one of the performance metrics being utilized. However, to access the usefulness of the predictor it is necessary to identify how close the prediction is from a reliable reference such as the native conformation. For this purpose, it is used the Root Mean Squared Deviation (RMSD), presented in Eq. (1), which can give a distance metric of how close the predicted conformation is from the native conformation. Then, it is possible to have a measure of how good the optimization is and how good the prediction is. This is very important since the improvement of one does not imply the improvement of the other. Since the works found in the literature which are compatible with this work only provided the RMSD, it is not possible to do a direct comparison with them using other metrics.

Four optimization models are employed, named SADE-DE-FRAG, SADE-DE-REMC, SADE-FRAG, and SADE-REMC, described in Sect. 4. For each variant of the proposed method, 10 independent runs were made for each of the target proteins. The tests were run on a machine equipped with an Intel® Core™ i5-3570k clocked at 4.2 GHz, 16 GB of RAM clocked at 1400 MHz, and running a GNU/Linux operating system.

The SADE-DE-FRAG and SADE-DE-REMC models use four fragment insertion operators each, plus six DE operators shown on Table 1. The SADE-FRAG and SADE-REMC models use only the four fragment insertion operators.

Results are compared with two works from the literature that are compatible with the proteins and metrics used in this work: MSA [22], which uses a multistage simulated annealing with fragment insertion, and the GA-APL [3], which uses a Genetic Algorithm with distribution of angles for each amino acid based on its predicted secondary structure.

All four proposed methods were run with a budget of 500000 function evaluations. This value was chosen to mimic the values used by the MSA algorithm. The parameters for all methods are exactly the same: population size is set to 100 and the learning phase of SaDE is set to 50 generations. The methods using fragment insertion can use up to 10 function evaluations per solution vector per generation with Monte Carlo criterion. This means that the number of generations can vary, due to a variable amount of function evaluation being spent with each fragment insertion operator. The average energy is compared using a Student's T-Test with a confidence of 95%.

6   Results and Analysis

Table 3 presents the results obtained. The first column shows the protein codename as in PDB, the second column presents the algorithm name, the third column shows the best (lowest) energy found in 10 runs, the fourth column shows the best (lowest) RMSD between the alpha carbons of the best-predicted protein and its native structure, and the fifth column shows the average energy and its standard deviation of 10 runs. The best values of each column, per protein, are marked in boldface. The average processing time was 7.43 ± 1.15, 10.34 ± 1.29, 11.87 ± 1.52 and 16.50 ± 1.58 min for 1ZDD, 1CRN, 1ENH, and 1AIL, respectively.

When comparing the minimum energy, RMSD and average energy of each algorithm, it is possible to notice that the methods proposed in this work are, in general, able to outperform both MSA and GA-APL. Considering only the proposed methods, both SADE-FRAG and SADE-REMC outperformed the other methods, with the exception of MSA on 1CRN when considering the average energy.

The comparison of the minimum energy reached within 10 runs of each algorithm shows that 3 of the 4 hybrid methods proposed achieved better energy values on the 1ZDD protein. For 1CRN, which is arguably the harder protein in this set, SADE-REMC achieved the best results and MSA scored the second best energy followed by SADE-FRAG. On the 1ENH protein, SADE-FRAG scored the best energy and SADE-REMC the second best, being apart by less than 2 units. For 1AIL protein, SADE-REMC got the best energy, more than ten units from the second best. With this information, it is possible to see that SADE-REMC is able to achieve the best energy values, or at least get a very close value to it.

Analyzing the RSMD it is possible to notice that for 1ZDD, near-native structures are found by all methods, with the exception of GA-APL. SADE-REMC achieved the best results for all proteins, with the exception of 1CRN,

Table 3. Results obtained

Protein	Algorithm	Min. energy	RMSDα(Å)	Avg. energy
1ZDD	GA-APL	−40.40	10.9	−36.20 ± 2.60
	MSA	−62.99	2.62	−48.96 ± 7.77
	SADE-DE-FRAG	−62.49	2.97	−30.44 ± 27.07
	SADE-DE-REMC	−70.06	1.81	−53.17 ± 11.33
	SADE-FRAG	−80.67	1.51	**−69.36 ± 5.50**
	SADE-REMC	**−82.46**	**1.16**	−68.36 ± 8.50
1CRN	GA-APL	−22.70	5.8	−18.20 ± 2.9
	MSA	−76.93	6.96	**−54.01 ± 17.30**
	SADE-DE-FRAG	−44.02	7.80	2.31 ± 55.94
	SADE-DE-REMC	−31.60	5.45	3.67 ± 36.82
	SADE-FRAG	−68.38	**5.38**	−45.29 ± 20.70
	SADE-REMC	**−82.72**	6.08	−23.18 ± 55.94
1ENH	GA-APL	−56.08	14.99	−51.52 ± 1.94
	MSA	−95.86	5.70	−80.75 ± 8.48
	SADE-DE-FRAG	−112.64	4.40	−87.63 ± 15.95
	SADE-DE-REMC	−113.93	3.90	−86.12 ± 30.66
	SADE-FRAG	**−127.31**	4.69	**−104.09 ± 12.06**
	SADE-REMC	−125.89	**3.23**	−98.75 ± 10.04
1AIL	GA-APL	−75.07	12.34	−71.08 ± 3.35
	MSA	−128.55	8.27	−117.54 ± 10.28
	SADE-DE-FRAG	−126.31	7.28	−96.70 ± 22.63
	SADE-DE-REMC	−149.37	6.43	−106.08 ± 31.94
	SADE-FRAG	−142.05	7.41	−118.39 ± 13.01
	SADE-REMC	**−159.48**	**4.46**	**−119.16 ± 25.01**

when it loses to SADE-FRAG. For 1CRN, all methods achieved similar results with a small amplitude between the methods, indicating the complexity of the protein. For 1ENH and 1AIL proteins, results with a correct global fold are found, however, lacking local alignments. This is confirmed later with visual analysis.

The average energy found during the experimentation indicates that SADE-FRAG and SADE-REMC achieved better results in all cases except for 1CRN. For 1CRN, as shown by the standard deviation, all four proposed models appear to have a varying result, while the competing methods had a smaller standard deviation. On 1ZDD and 1ENH proteins, SADE-FRAG and SADE-REMC have statistically equal results and are able to outperform all the others as confirmed with an unpaired Student's T-Test. In the 1CRN protein, SADE-REMC, SADE-FRAG and MSA are all statistically equal. However, MSA has a smaller mean and standard deviation. As for 1AIL, all four proposed methods and MSA are equivalent.

Table 4. Results showing the GDT-TS metric

Protein	SADE-FRAG	SADE-REMC
	$(Best)Avg \pm StdDev$	$(Best)Avg \pm StdDev$
1ZDD	$(0.860)0.779 \pm 0.072$	$(0.860)0.762 \pm 0.066$
1CRN	$(0.543)0.422 \pm 0.057$	$(0.462)0.405 \pm 0.047$
1ENH	$(0.463)0.414 \pm 0.036$	$(0.463)0.390 \pm 0.037$
1AIL	$(0.407)0.342 \pm 0.031$	$(0.449)0.365 \pm 0.039$

Table 5. Results showing the TM-Score metric

Protein	SADE-FRAG	SADE-REMC
	$(Best)Avg \pm StdDev$	$(Best)Avg \pm StdDev$
1ZDD	$(0.658)0.547 \pm 0.091$	$(0.666)0.542 \pm 0.085$
1CRN	$(0.393)0.288 \pm 0.041$	$(0.295)0.261 \pm 0.033$
1ENH	$(0.288)0.254 \pm 0.022$	$(0.292)0.237 \pm 0.025$
1AIL	$(0.338)0.281 \pm 0.030$	$(0.440)0.298 \pm 0.051$

Tables 4 and 5 present for the two best methods, SADE-REMC and SADE-FRAG, their respective GDT-TS and TM-Score. It can be seen that for 1ZDD the mean value of the predictions was above 0.5, meaning that both algorithms got the same fold. 1CRN and 1ENH proteins had similar results for GDT-TS, with a value close to 0.4, indicating that the method was close to arriving at a solution in the same fold. Proteins 1CRN, 1ENH and 1AIL had similar TM-Scores of about 0.28, showing that it started to converge but did not get to the same fold as the native structure. With this information, it is possible to see that there is more room for improvement than just the one pointed by RMSD. There is a strong indication that the method's performance decreases as the number of amino acids increases, which was expected since the number of variables grows making the problem more complex. Another aspect worth noting is that GDT-TS and TM-Score do agree on which method was the best considering the mean. However, regarding the relative performance between two proteins, the metrics do not always agree. For instance, TM-Score points out that SADE-FRAG was better on 1AIL than on 1ENH, but GDT-TS points otherwise. This happens due to different aspects being measured by each metric.

With the intent of a visual evaluation of the results obtained, the conformations with best RMSDs, selected from Table 3, were plotted in Fig. 2 with the predicted protein in cyan/light color and the native conformation in red/dark color. The 1ZDD protein achieved the best overall RMSD and is shown in Fig. 2(a).

It is possible to notice only small deviations from the native conformation. The 1CRN protein, shown in Fig. 2(b), is the hardest target protein approached in this work, and this can be confirmed by the visual analysis. The shape of the fold is mostly correct on the prediction, with one α-helix rotated approximately 130° from where it should be. The main loop of the protein also is close to the native conformation. However, the β-sheets were not found by the prediction. On 1ENH protein, Fig. 2(c), the prediction was very accurate in most part of the protein, however at both ends the α-helices were misfolded. For 1AIL protein, Fig. 2(d), the analysis is similar to the last protein. Two of the three α-helices present in this protein are correctly predicted with one of them missing a single turn, and the other one lacking 1 turn at the beginning and approximately 3 turns at the end. This is responsible for most of the error found in the prediction of this protein.

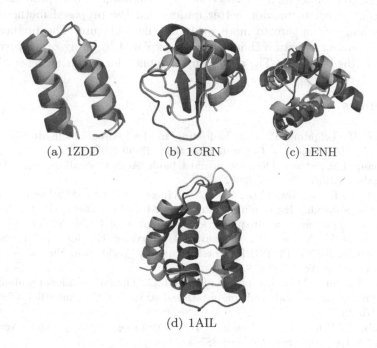

(a) 1ZDD   (b) 1CRN   (c) 1ENH

(d) 1AIL

Fig. 2. Predicted proteins (cyan, lighter) compared to the native conformations (red, darker) (Color figure online)

7   Conclusions and Future Work

This work presented four hybrid methods for solving the PSPP based on SaDE and non-homologous fragment insertion guided by the Monte Carlo algorithm. In our application, the initial population is generated with a Monte Carlo phase using a relaxed score function and a repacking stage is employed at the end of the optimization. Two proposed approaches, SADE-REMC and SADE-FRAG, were both able to outperform other algorithms found in the literature. In light of the experimentation and its results, it was observed that the hybridization of the SaDE algorithm with domain-specific operators achieved better results. Also, the use of fragment insertion leads to a better result than methods guided only by using the energy function. The visual analysis of the best conformations showed that in 1ENH and 1AIL proteins the main source of error was due to incorrect prediction of the secondary structure.

Furthermore, the use of GDT-TS and TM-Score metrics allowed to gather more insights about the performance of the proposed methods. With that in mind, it was presented the measurements in the hope that it would be used as a reference for future works. Also, the use of better metrics than the RMSD start to become more frequently used.

As future research directions, it is possible to explore other self-adaptive schemes for controlling fragment insertion. The inclusion of more proteins should also be considered, to provide better understand the proposed method's performance and to compare to more works in the literature. The prediction of secondary protein structure should also be explored by using other predictors available in the literature. The same can be applied for fragment generation.

References

1. Berger, B., Leighton, T.: Protein folding in the hydrophobic-hydrophilic (HP) model is NP-complete. J. Comput. Biol. **5**(1), 27–40 (1998)
2. Berman, H.M., et al.: The protein data bank. Acta Crystallogr. Sect. D: Biol. Crystallogr. **58**(6), 899–907 (2002)
3. Borguesan, B., e Silva, M.B., Grisci, B., Inostroza-Ponta, M., Dorn, M.: APL: an angle probability list to improve knowledge-based metaheuristics for the three-dimensional protein structure prediction. Comput. Biol. Chem. **59**, 142–157 (2015)
4. Buchan, D.W., Minneci, F., Nugent, T.C., Bryson, K., Jones, D.T.: Scalable web services for the PSIPRED Protein Analysis Workbench. Nucleic Acids Res. **41**(W1), W349–W357 (2013)
5. Dorn, M., e Silva, M.B., Buriol, L.S., Lamb, L.C.: Three-dimensional protein structure prediction: methods and computational strategies. Comput. Biol. Chem. **53**, 251–276 (2014)
6. Drenth, J.: Principles of Protein X-Ray Crystallography. Springer, New York (2007). https://doi.org/10.1007/0-387-33746-6
7. Garza-Fabre, M., Kandathil, S.M., Handl, J., Knowles, J., Lovell, S.C.: Generating, maintaining, and exploiting diversity in a memetic algorithm for protein structure prediction. Evol. Comput. **24**(4), 577–607 (2016)
8. Habeck, M., Nilges, M., Rieping, W.: Replica-exchange Monte Carlo scheme for Bayesian data analysis. Phys. Rev. Lett. **94**(1), 018105 (2005)
9. Hart, W.E., Istrail, S.: Robust proofs of NP-hardness for protein folding: general lattices and energy potentials. J. Comput. Biol. **4**(1), 1–22 (1997)
10. Karafotias, G., Hoogendoorn, M., Eiben, Á.E.: Parameter control in evolutionary algorithms: trends and challenges. IEEE Trans. Evol. Comput. **19**(2), 167–187 (2015)

11. Liu, J., Li, G., Yu, J., Yao, Y.: Heuristic energy landscape paving for protein folding problem in the three-dimensional HP lattice model. Comput. Biol. Chem. **38**, 17–26 (2012)
12. Lopes, H.S.: Evolutionary algorithms for the protein folding problem: a review and current trends. In: Smolinski, T.G., Milanova, M.G., Hassanien, A.E. (eds.) Computational Intelligence in Biomedicine and Bioinformatics. SCI, vol. 151, pp. 297–315. Springer, Heidelberg (2008). https://doi.org/10.1007/978-3-540-70778-3_12
13. Moult, J., Fidelis, K., Kryshtafovych, A., Schwede, T., Tramontano, A.: Critical assessment of methods of protein structure prediction: progress and new directions in round XI. Proteins: Struct. Funct. Bioinf. **84**(S1), 4–14 (2016)
14. Narloch, P.H., Parpinelli, R.S.: The protein structure prediction problem approached by a cascade differential evolution algorithm using ROSETTA, pp. 294–299. IEEE (2017)
15. Nunes, L.F., Galvão, L.C., Lopes, H.S., Moscato, P., Berretta, R.: An integer programming model for protein structure prediction using the 3D-HP side chain model. Discrete Appl. Math. **198**, 206–214 (2016)
16. de Oliveira, S.H., Shi, J., Deane, C.M.: Building a better fragment library for de novo protein structure prediction. PloS One **10**(4), e0123998 (2015)
17. Parpinelli, R.S., Lopes, H.S.: New inspirations in swarm intelligence; a survey. Int. J. Bio-Inspir. Comput. **3**(1), 1–16 (2011)
18. Parpinelli, R.S., Plichoski, G.F., Da Silva, R.S., Narloch, P.H.: A review of technique for on-line control of parameters in swarm intelligence and evolutionary computation algorithms. Int. J. Bio-Inspir. Comput. (IJBIC) (2019, accepted for publication)
19. Price, K., Storn, R.M., Lampinen, J.A.: Differential Evolution: A Practical Approach to Global Optimization. Springer, Heidelberg (2005). https://doi.org/10.1007/3-540-31306-0
20. Qin, A.K., Huang, V.L., Suganthan, P.N.: Differential evolution algorithm with strategy adaptation for global numerical optimization. IEEE Trans. Evol. Comput. **13**(2), 398–417 (2009)
21. Qin, A.K., Suganthan, P.N.: Self-adaptive differential evolution algorithm for numerical optimization. In: The 2005 IEEE Congress on Evolutionary Computation, vol. 2, pp. 1785–1791. IEEE (2005)
22. Silva, R.S., Parpinelli, R.S.: A multistage simulated annealing for protein structure prediction using Rosetta. In: Anais do Computer on the Beach, pp. 850–859 (2018)
23. Vanderbilt, D., Louie, S.G.: A Monte Carlo simulated annealing approach to optimization over continuous variables. J. Comput. Phys. **56**(2), 259–271 (1984)
24. Walsh, G.: Proteins: Biochemistry and Biotechnology. Wiley, Hoboken (2002)
25. Wang, Y., Cai, Z., Zhang, Q.: Differential evolution with composite trial vector generation strategies and control parameters. IEEE Trans. Evol. Comput. **15**(1), 55–66 (2011)
26. Xu, J., Zhang, Y.: How significant is a protein structure similarity with TM-score = 0.5? Bioinformatics **26**(7), 889–895 (2010)
27. Zemla, A.: LGA: a method for finding 3D similarities in protein structures. Nucleic Acids Res. **31**(13), 3370–3374 (2003)
28. Zhang, J., Sanderson, A.C.: JADE: adaptive differential evolution with optional external archive. IEEE Trans. Evol. Comput. **13**(5), 945–958 (2009)
29. Zhang, X., et al.: 3D protein structure prediction with genetic tabu search algorithm. BMC Syst. Biol. **4**(1), S6 (2010)
30. Zhang, Y., Skolnick, J.: Scoring function for automated assessment of protein structure template quality. Proteins: Struct. Funct. Bioinf. **57**(4), 702–710 (2004)

Scheduling Simultaneous Resources: A Case Study on a Calibration Laboratory

Roberto Tavares Neto$^{(\boxtimes)}$ (iD) and Fabio Molina da Silva$^{(\boxtimes)}$ (iD)

Federal University of Sao Carlos, Sao Paulo, Brazil
{tavares,fabio}@dep.ufscar.br

Abstract. A calibration laboratory studied in this research performs a thermal test that requires an analyst for setup and processing and an oven to perform such an essay. For convenience, it's possible to group some of the essays according to the oven capacity. In this scenario, this paper proposes a scheduling approach to minimize the total flowtime of the orders. This is a multiple resource scheduling problem, where a resource (operator) is used on two processes (oven setup and analysis). In contrast to the classical definition of multiple resource scheduling problems, the oven setup process requires the presence of the operator only for the startup of the process. To solve this problem, we derived: (i) a mixed-integer formulation; (ii) an Ant Colony Optimization (ACO) approach. On those developments, we also discuss some structural properties of this problem, that may lead to further advances in this field in the future. Our results show the ACO approach as a good alternative to the MIP, especially when solving instances with 30 service orders.

Keywords: Multiple resource scheduling · Ant Colony Optimization
Multiple constraint scheduling

1 Introduction

From time to time, pieces of equipment used by manufacturing and service systems require to be submitted to some calibration procedure. This paper approaches a problem found by a laboratory that serves mainly an aircraft service facility. In this problem, there is a set of calibration service orders on a process that uses two resources: a human analyst and an oven. The human analyst is required to program the oven and to elaborate the report of each piece. The oven is used to perform the essays after the programming. It's possible to group the essays in batches, respecting the oven capacity. The reports, on the other hand, must be processed individually. The goal of the laboratory is to find

This research was supported by CAPES, CNPq (407104/2016-0) and FAPESP (2010/10133-0).

the processing sequence of the activities of both the analyst and the oven that minimizes the system flowtime.

Formally, this problem has some characteristics already addressed by the scientific literature (e.g., [1]). It's clear that we are dealing with a 2-stage flowshop manufacturing environment. The sequence of operations is not required to be the same on the two stages, making this problem close to a non-permutation flowshop environment. On this flowshop, batches are allowed on the first stage.

Another feature of this problem is the requirement of two resources to be allocated to a single operation (the oven processing). Some studies have addressed the issue of multiple resources scheduling. In those papers, all the resources allocated to operate are used during all the processing, being released together at the end of it (a comprehensive review on the topic can be found in [21]). This constraint does not apply to the problem studied in this paper: in this case, there is an operation that requires two resources (analyst and oven) to be started. Furthermore, after being released, the "analyst resource" can be allocated to work on unfinished reports while the oven is performing the rest of their operation.

To address this problem, the remaining of this paper is organized as follows: Sect. 2 formally defines the problem; Sect. 3 presents some concepts from the literature regarding the above-mentioned problems; Sect. 4.1 presents a mathematical formulation of the problem; Sect. 4.2 presents a heuristic approach for the problem; furthermore, Sect. 5 presents the results of the solution methods on a large set of problem instances; finally, Sect. 6 presents the final remarks of this paper.

2   Problem Definition

As stated previously, this paper tries to solve a scheduling problem found on a calibration laboratory with two processing resources: an oven and an analyst. The analyst is required to perform the initial setup of the oven, and then is released to perform another duties. Each processing order requires two sequential operations: the essay (performed by the oven) and the analysis (performed by the analyst). Several essays can be grouped, according to the oven capacity. The objective is to minimize the total system flowtime. Figure 1 presents a possible three-order scheduling. In this example, the essay of orders 2 and 3 are grouped and executed at $t = 0$. Once the essays are finished, it is performed the analysis of order 2. When the analysis of order 2 is finished, the analyst is released to perform the setup for the essay of order 1. After the setup is finished, the analyst perform the analysis of order 3 while the oven is processing the essay of order 1. When the essay of order 1 is completed, the analyst perform the analysis of order 1.

To represent this problem, this paper groups the activities into batches. A batch of activities is composed by all operations performed between two oven operations. Thus, in Fig. 1, the first batch ($b = 0$) is composed by oven operation [2+3], the corresponding analyst setup and the reporting operation [2]. Similarly, the second batch is composed by the oven/analyst setup operation of [1] and the reporting operation of [3, 1].

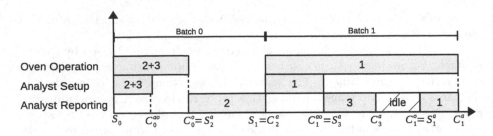

Fig. 1. A simple scheduling of three service orders grouped into two batches

Table 1. Symbols used on this formulation

Symbol	Type	Description
N	Set	Set of orders
S	Set	Set of all subset of vertices of the graph represented by y_{ijb}
ρ_{0i}	Real	Processing time for the oven operation of order i
ρ_{1i}	Real	Time required to perform the oven setup by the analyst
ρ_{2i}	Real	Time required to perform analysis on order i
b	Integer	Index for setup batch task
$\{i, j, k\}$	Integer	Indexes for an analysis task (0 is a dummy order)
B	Integer	Max number of elements allowed in a single batch
x_{ib}	Binary	Assumes 1 if order i is allocated into batch b, 0 otherwise
y_{ijb}	Binary	Assumes 1 if analysis j is performed after analysis i at batch b 0 otherwise
S_b	Real	Starting time of batch b
C_b^o	Real	Completion time of operation performed by the oven at batch b
C_b^{ao}	Real	Completion time of operation performed by the operator to execute the oven setup at batch b
S_i^a	Real	Starting time of reporting operation of order i
C_i^a	Real	Completion time of reporting operation of order i

Formally, this paper adopts a set of symbols to address each element of this problem and the corresponding solution. Those symbols are presented in Table 1.

3   Literature Overview

3.1   Multiple Resource Scheduling

The coordination between limited resources is also a key factor in healthcare related problems. As an example, [22] presents a branch-and-price approach and apply it to two case studies of scheduling patients on exam labs, coordinating the assignment of exam rooms and medical personnel. In this problem, besides the need for simultaneous allocation between the physical and human resources, there is a no-wait constrain imposed. Likewise, manufacturing facilities demand a careful assignment of limited resources. This can be found on injection molding operations (e.g., see [7]) and semiconductor manufacturing (e.g., [12]). The class of problems that allows the assignment of both machines and workers is usually referred as Dual Resource Constrained (DRC) systems. Several solution methods have been proposed in the literature: e.g., [13] uses a Branch Population Genetic Algorithm (BPGA) to address a DRC problem consisting of allocating machines and workers to process parts of manufacturing jobs. Further heuristic and metaheuristic-based approaches for the DRC systems can be found in the literature (e.g., [20]).

All the researches mentioned allocates the multiple resources during all the processing time of the operation. However, in the specific case of the problem approached by this paper, the oven and analyst resources are jointly allocated only on the initial phase of the oven operation.

3.2 Batching in Two-Stage Flowshops

The relevance of batching to real-world applications is well-documented in the literature (e.g., see [3]). This category of problems has been solved using a wide range of techniques: [14] presents a mathematical formulation for a scheduling problem considering a flowshop when batches are formed at the first stage and then processed by all stages. Further formulations are presented by [2], among others. Further approaches can also be found: e.g., [19] proposes a dynamic programming algorithm; [23] uses a genetic algorithm.

All of those researches considered single-resource scheduling problems. Unfortunately, we could not find a solution for a batching flowshop environment with the characteristics considered by this paper.

3.3 The Ant Colony Optimization

The Ant Colony Optimization [8,9,11] is a nature-inspired meta-heuristic that creates solutions based on the movement of virtual agents (named "ants") on a graph. Those agents mimic the behavior of real ants use to gather food to the nest: initially, the ants move randomly, based only on the neighborhood characteristics. When some food source is found, the ant return to the nest depositing a chemical substance named *pheromone*. This substance slowly evaporates after its release. Once an ant can perceive any pheromone trail, the chance of choosing routes with higher pheromone values increases. After some ants perform this cycle, the pheromone levels related to the path elected by most ants became dominant. Thus, the next ants choose to use this path to reach the food source. Figure 2 represents the mechanism of pheromone updating.

To implement this behavior, the literature (e.g., see [8–11]) presents some common approaches:

- The pheromone is represented as a τ_{ij} array, representing the value of the pheromone on the arc $i \rightarrow j$;

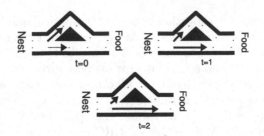

Fig. 2. The pheromone evolution during an ACO algorithm execution

- The τ_{ij} array is updating according Eq. 1 ($\Delta\tau$ is zero if $i \to j$ does not belong to the current solution).
- The neighborhood characteristics are modeled by a **visibility function** η_{ij}. This function is usually found by some problem-specific characteristics (e.g., when solving the traveling salesman problem, it's usual to describe the visibility as a function of the distances between two cities).
- The next movement of the ant is given by a probabilistic function weighted according to a **transition rule** presented in Eq. 2.
- γ, β and $\Delta\tau$ are parameters.

$$\tau_{ij}(t) = \gamma \cdot \tau_{ij}(t-1) + \Delta\tau \tag{1}$$

$$P_{ij} = \frac{\tau_{ij} \cdot \eta_{ij}^{\beta}}{\displaystyle\sum_{k=0}^{max(i)} \tau_{kj} \cdot \eta_{kj}^{\beta}} \tag{2}$$

Formally, the pseudo-code for an ACO algorithm is presented in Algorithm 1.

Algorithm 1: The ACO pseudo-code

---

1 Initialize;
2 **repeat** *At this level, each execution is called* ***iteration***
3   Each ant is positioned on the initial node;
4   **repeat** *At this level, each execution is called* ***step***
5   Each ant applies a state transition rule to increment the solution;
6   Apply the pheromone local update rule;
7   **until** *all the ants have built a complete solution*;
8   Apply a local search procedure;
9   Apply the pheromone global update rule;
10 **until** *the stop criteria is satisfied*;

Several researches use ACO-based algorithms to tackle important scheduling problems. E.g., [17] presents an ACO algorithm to schedule orders into a flow-shop; [4] combines ACO and Bean Search to generate schedules for an open-shop environment; [16] uses an ACO algorithm to optimize a project schedule. Mostly problem structures of those implementations uses a full-connected graph. As will be presented in Sect. 4.2, this paper uses a different representation, more suitable to a pure assignment problem.

4 Solution Strategies

Two approaches were developed to solve the problem presented in this paper: a mixed-integer programming (MIP) model and an ACO algorithm. Although the MIP could, in theory, delivery the optimal solution, the computational effort require is usually prohibited. This strategy is used in this paper to validate the results of the ACO and to obtain optimal solutions for the small-sized problems.

4.1 A Mixed-Integer Formulation

The first approach for this problem on this paper was a Mixed-Integer Programming model. This approach was used to: (i) assist to formally define the problem; and (ii) obtain optimal results to allow the assessment of the algorithm proposed in Sect. 4.2.

Equations 3–5 guarantees the validity of the setup batches formed by the model. Equations 3 assures that any non-dummy order must be assigned just once. The oven capacity is constrained by Eq. 4. A dummy order is assigned to all setup batches, as stated by Eq. 5.

$$\sum_{0 \leq b < |N|} x_{ib} = 1 \qquad\qquad \forall \left\{ 0 < i < |N| \right. \tag{3}$$

$$\sum_{0 \leq b < |N|} x_{ib} \leq B \qquad\qquad \forall \left\{ 0 \leq b < |N| \right. \tag{4}$$

$$x_{0b} = 1 \qquad\qquad \forall \left\{ 0 \leq b < |N| \right. \tag{5}$$

Equations 6–12 grant the validity of the solution regarding the y_{ijk} variables. Those constrains are common in the VRP literature (e.g., see [6]), and can be divided in two subgroups: Eq. 6 assures that an order i cannot be scheduled after itself; Eqs. 7–9 assures that each order is performed in just one batch; Eqs. 10–11 assures that all schedules start at the dummy node; Eq. 12 guarantee the structure of a schedule on each batch.

$$y_{iib} = 0 \qquad\qquad \forall \begin{cases} 1 \leq i < |N| \\ 0 \leq b < |N| \end{cases} \tag{6}$$

$$\sum_{\substack{1 \leq i < |N| \\ 0 \leq b < |N|}} y_{ijb} = 1 \qquad\qquad \forall \left\{ 1 \leq j < |N| \right. \tag{7}$$

$$\sum_{\substack{1\le j<|N| \\ 0\le b<|N|}} y_{ijb} = 1 \qquad\qquad \forall\{1\le i<|N| \tag{8}$$

$$\sum_{0\le i<|N|} y_{kjb} = \sum_{0\le i<|N|} y_{ikb} \qquad\qquad \forall\begin{cases} 0\le k<|N| \\ 0\le b<|N| \end{cases} \tag{9}$$

$$\sum_{0\le i<|N|} y_{i0b} = 1 \qquad\qquad \forall\{0<b<|N| \tag{10}$$

$$\sum_{0\le j<|N|} y_{0jb} = 1 \qquad\qquad \forall\{0\le b<|N| \tag{11}$$

$$\sum_{\{i,j\}\in S} x_{ijb} \le |S| - 1 \qquad\qquad \forall\begin{cases} 0\le b<|N| \\ S\subset N \\ 2\le |S|\le |N|/2 \end{cases} \tag{12}$$

The starting and completion times are found by Eqs. 13–22, widely present in the Vehicle Routing Problem literature (VRP - e.g., see [5,6]). Equations 13–16 state that a batch other than the first one can only be processed when: 14–15 the previous setup has finished at the oven and analyst; and 16 any previous reporting activity has finished. Equations 17–18 establish the processing times for each batch. Equations 19–22 determine the values of the starting and completion times of the reporting orders.

$$S_0 = 0 \tag{13}$$

$$S_b \ge C^o_{(b-1)} \qquad\qquad \forall\{0<b<|N| \tag{14}$$

$$S_b \ge C^{ao}_{(b-1)} \qquad\qquad \forall\{0<b<|N| \tag{15}$$

$$S_b \ge C^a_i - M\cdot\Big(1 - \sum_{1\le j<|N|} y_{ijb}\Big) \qquad\qquad \forall\{0<b<|N| \tag{16}$$

$$C^o_b \ge S_b + \rho_{0,i}\cdot x_{ib} \qquad\qquad \forall\begin{cases} 0\le b<|N| \\ 1\le i<|N| \end{cases} \tag{17}$$

$$C^{ao}_b \ge S_b + \rho_{1,i}\cdot x_{ib} \qquad\qquad \forall\begin{cases} 0\le b<|N| \\ 1\le i<|N| \end{cases} \tag{18}$$

$$S^a_i \ge C^o_b - M\cdot(1 - x_{ib}) \qquad\qquad \forall\begin{cases} 0\le b<|N| \\ 1\le i<|N| \end{cases} \tag{19}$$

$$S^a_i \ge C^{ao}_b - M\cdot(1 - x_{ib}) \qquad\qquad \forall\begin{cases} 0\le b<|N| \\ 1\le i<|N| \end{cases} \tag{20}$$

$$S^a_j \ge C^a_i - M\cdot(1 - y_{ijb}) \qquad\qquad \forall\begin{cases} 0\le b<|N| \\ 1\le i<|N| \\ 1\le j<|N| \end{cases} \tag{21}$$

$$C^a_i = S^a_i + \rho_{2,i} \qquad\qquad \forall\{1\le i<|N| \tag{22}$$

4.2   An Ant Colony Optimization Approach

Two main design choices were made to allow the application of the concepts of the ACO meta-heuristic presented in Sect. 3.3 on this problem: the graph encoding and a corresponding fitness function.

Graph Modeling. As presented in Sect. 3.3, ACO algorithms require a graph that allows the ant to create a solution. In the specific case of this implementation the graph is composed of two stages: the first stage allows the ant to allocate orders into groups of oven operations; the second stage allows the ant to allocate reporting activities into batches, following the concept of batches presented in Sect. 2. Since the order of the assignment is not important in this data structure, it's possible to adopt a directed graph as shown in Fig. 3[1]. On this graph, there are two groups of vertices. On each group, there is one line for each batch and one column for each order. The left group represents the allocation of oven operations into batches, and the right represents the allocation of reporting operations. The ant builds a new solution by moving from left to right. Figure 3 indicates how an ant would use such graph to build the solution presented in Fig. 1.

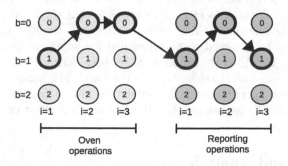

Fig. 3. A graph representing a 3-orders, 3-batches problem. The solution highlighted is the one presented in Fig. 1

The Visibility Function. Two different visibility functions were defined, according the choice to be made by the ant. When moving through the oven operations part of the graph, the ant uses the visibility presented in Eq. 23. On the case of reporting operations, the algorithm adopts Eq. 24. On this equation, b' is the batch that executes the oven operations of order i.

$$\eta_{ib} = \begin{cases} \frac{1}{1+|\max_{j \in b}\{x_{jb}\} - \rho_{0i}|} & \text{if } b \text{ is the current batch} \\ \frac{1}{1+\rho_{0i}} & \text{otherwise} \end{cases} \qquad (23)$$

$$\eta_{ib} = \begin{cases} 1 & \text{if } b \geq b' \\ 0 & \text{otherwise} \end{cases} \qquad (24)$$

---

[1] For convenience, the dummy node $i = 0$ is not represented here.

The Fitness Function. Given a batch assignment for each analysis, each value of C_i^a can be found by using Algorithm 2. On this algorithm, for each batch b, the orders $i \in b$ are divided into two sets: Π_a, containing all orders processed by the oven on previous batches (and can be executed at any moment); and Π_b, containing all orders which oven operations are executed on the current batch (and must wait until C_b^o to start the processing). In the example of Fig. 1, when $b = 1$, $\Pi_a = \{3\}$ and $\Pi_b = \{1\}$. This algorithm is composed of three different stages:

STAGE 1. Orders Π_a by ρ_{2i} and schedule them until the idle time is not reached.

STAGE 2. If there is at least a non-scheduled order remaining at Π_a, decides if this order must use the idle time or be delayed.

STAGE 3. All the remaining orders are sorted by ρ_{2i} and then scheduled using this sequence.

The optimal scheduling of orders at Stage 1 and Stage 3 is straightforward: since it's well-known that the minimum flowtime of a single processor is achieved by ordering the orders by their processing time [18]. Thus, the optimal scheduling of Stage 1 and 3 is achieve by ordering the jobs according the values of ρ_{2i}.

When scheduling the jobs of Π_a, it's possible that the total processing time is higher than the idle time of the batch. In this case, the first job that violates the idle time can be assigned to be scheduled following the previous sequence of Π_a or moved to Π_b. One can show that, considering ρ_a as the processing time of the first job i of Π_a that violates the batch idle time and ρ_b the lower processing time of all batches of Π_b, i must be moved from Π_a to Π_b only if $\rho_a - \rho_b > 2 \cdot Idle$, $Idle$ represents the idle time of the analyst on this processing batch.

5 Results and Analysis

The model presented in Sect. 4.1 was coded in Python 3 with CPLEX 12.8 library. It was stated a maximum running time of 3600 s. The ACO algorithm presented in Sect. 4.2 was coded in C++, and compiled using GCC 4.8.4. The IRACE package [15] was used to obtain the optimal values of the parameters of ACO algorithm. The following parameters were considered: number of cycles: 100, 1000 or 5000; initial value of the pheromone: 100 or 200; pheromone evaporation constant: 0.8, 0.85, 0.9, 0.95, 0.99 or 1; pheromone increase constant: 0, 1, 2, 3, 4 or 5; β: 1, 2, 3, 4, or 5. For each class of problem, one instance was randomly selected and the resulting set of instances were used by IRACE to perform the tunning. All the remaining configurations were set as the default values. After running the IRACE package, the optimal parameter setting was: 100 cycles; initial value of pheromone: 100; pheromone evaporation constant: 0.8; pheromone increase constant: 5; β: 1.

A set of 960 instance files were generated. The parameters used to generate the files, defined based on information gathered on the calibration laboratory, are presented in Table 2.

Algorithm 2: The pseudo-code for finding the final fitness and the sequence of reporting activities given an ACO solution.

1 $\Pi_a \leftarrow$ set of jobs that can be started at any moment;
2 $\Pi_b \leftarrow$ set of jobs that can be started only after the oven operation is finished;
3 $\Pi_s \leftarrow$ final sequence;
4 $Idle \leftarrow$ The idle time of the analyst;
5 $C_i \leftarrow$ The Completion Time of order i;
6 **begin**
7   **foreach** *batch* k **do**
8   Order Π_a and Π_b by the values of $\rho_{2,i}$;
9          $\Pi_s \leftarrow \emptyset$;
10   **while** $\Pi_a \neq \emptyset$ *and* $\rho_{2,\Pi_{a[0]}} + \sum_i \rho_{2,\Pi_{b[i]}} \leq Idle$ **do**
11             $\lfloor$ Move $\Pi_{a[0]}$ to the end of Π_s;
12   **if** $\Pi_a \neq \emptyset$ **then**
13   **if** $\rho_{(2,\Pi_{a[0]})} + \rho_{2,\Pi_{b[0]}} \leq 2 \cdot Idle$ **then**
14                 $\lfloor$ Move $\Pi_a[0]$ to the end of Π_s;
15         $C_{\Pi_s[0]} = C^a o_k + \rho_{2,\Pi_s[0]}$;
16         $C_{\Pi_s[i]} = C_{\Pi_s[i-1]} + \rho_{2,\Pi_s[i]}, \forall 0 < i \leq |\Pi_s|$;
17         $\Pi_b \leftarrow \Pi_b \bigcup \Pi_a$;
18   Order Π_b by the values of $\rho_{2,i}$;
19         $C_{\Pi_b[0]} = max_i\{C_{\Pi_s[i]}\} + \rho_{2,\Pi_b[0]}$;
20         $C_{\Pi_b[i]} = C_{\Pi_b[i-1]} + \rho_{2,\Pi_b[i]}, \forall 0 < i \leq |\Pi_b|$;
21   Append Π_b at the end of Π_s;
22 **return** $\Pi_s, \sum_i C_i$

Table 2. Parameters used to generate the instance files

Parameter	Values
Number of files	20 for each combination of $|N|$ and B
$|N|$	5, 10, 15, 20, 30, 40, 50, 100 or 200 orders
B	4, 6 or 8 orders
ρ_{0i}	Integer sampled from $[106, 345]$
ρ_{1i}	Real number sampled from $[0.2, 0.4] \cdot \rho_{0i}$
ρ_{2i}	Integer sampled from $[20, 75]$

The first analysis performed is regarding the fitness found by both algorithms. To perform this comparison, for each fitness measure, it was calculated the Gap, $Gap = (\text{``}FitnessFound\text{''} - \text{``}BestFitness\text{''})/\text{``}BestFitness\text{''}$. A graphical and a descriptive statistical analysis are presented in Fig. 4 and Table 3. As presented, the exact approach could obtain better results on mostly problem instances containing 5, 10 and 20 orders. For larger problems, the mathematical model could not reach the best value and ACO presented better results.

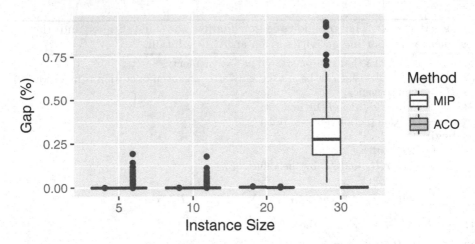

Fig. 4. Values of Gap found, according to the instance size and method

Table 3. Descriptive statistics analysis for the fitness results

Method	Size	Min	Average	Std. Dev.	Max
MIP	5	0.00	0.00	0.00	0.00
	10	0.00	0.00	0.00	0.00
	20	0.00	0.00	0.00	0.01
	30	0.03	0.30	0.17	0.95
ACO	5	0.00	0.01	0.02	0.19
	10	0.00	0.01	0.02	0.18
	20	0.00	0.00	0.00	0.00
	30	0.00	0.00	0.00	0.00

Table 4 presents the time required to run each algorithm according to the instance sizes. As expected, the ACO approach demands less resources than he MIP approach.

Given those results, some analysis can be derived:

For small-sized instances (5, 10 and 20), the MIP approach was able to obtain the optimal solution within the time limit imposed. For instances of size 5 and 10, the ACO algorithm behavior was slightly worse, with more outliers and a standard deviation of 0.2. For instances of size 20, the fitness found by both approaches were practically the same.

On instances of size 30, the MIP approach were not able to reach the optimal solution on any instance. On those cases, the ACO presented a superior behavior.

Table 4. Descriptive statistics analysis for the running time

Method	Size	Min	Average	Std. Dev.	Max
MIP	5	0.00	0.27	0.72	5.00
	10	3600.00	3602.03	4.42	3628.00
	20	3600.00	3601.99	3.75	3631.00
	30	3600.00	3601.44	3.35	3624.00
ACO	5	0.00	0.00	0.00	0.00
	10	0.01	0.01	0.00	0.01
	20	0.02	0.03	0.01	0.04
	30	0.03	0.05	0.01	0.07

6 Final Remarks

This paper approaches the problem of scheduling thermical essays of a laboratory. In this problem, an analyst is require to perform a setup of the oven and to elaborate reports. The oven is occupied during the setup and execution of the essay.

To solve this problem, this paper proposed a representation scheme and two solutions methods - a MIP formulation and an ACO algorithm. It was clear that the ACO algorithm could obtain good results requiring significantly less resources than the MIP.

In further works, are planned the development of more specific exact methods (since the pure MIP model does not seems to be adequate to solve instances larger than 30 orders) and local search methods to be used by the ACO.

References

1. Allahverdi, A., Ng, C., Cheng, T., Kovalyov, M.Y.: A survey of scheduling problems with setup times or costs. Eur. J. Oper. Res. **187**(3), 985–1032 (2008). https://doi.org/10.1016/j.ejor.2006.06.060. http://www.sciencedirect.com/science/article/pii/S0377221706008174
2. Behnamian, J., Ghomi, S.F., Jolai, F., Amirtaheri, O.: Realistic two-stage flow-shop batch scheduling problems with transportation capacity and times. Appl. Math. Model. **36**(2), 723–735 (2012). https://doi.org/10.1016/j.apm.2011.07.011. http://www.sciencedirect.com/science/article/pii/S0307904X1100391X
3. Belaid, R., Tkindt, V., Esswein, C.: Scheduling batches in flow-shop with limited buffers in the shampoo industry. Eur. J. Oper. Res. **223**(2), 560–572 (2012). https://doi.org/10.1016/j.ejor.2012.06.035. http://www.sciencedirect.com/science/article/pii/S0377221712004900
4. Blum, C.: Beam-ACO-hybridizing ant colony optimization with beam search: an application to open shop scheduling. Comput. Oper. Res. **32**(6), 1565–1591 (2005)
5. Braekers, K., Ramaekers, K., Nieuwenhuyse, I.V.: The vehicle routing problem: state of the art classification and review. Comput. Ind. Eng. **99**, 300–313 (2016). https://doi.org/10.1016/j.cie.2015.12.007. http://www.sciencedirect.com/science/article/pii/S0360835215004775

6. Cordeau, J.F., Laporte, G., Savelsbergh, M.W., Vigo, D.: Vehicle routing. In: Barnhart, C., Laporte, G. (eds.) Handbooks in Operations Research and Management Science: Transportation, vol. 14, pp. 367–428. Elsevier, Amsterdam (2007). https://doi.org/10.1016/S0927-0507(06)14006-2. http://www.sciencedirect.com/science/article/pii/S0927050706140062

7. Dastidar, S.G., Nagi, R.: Scheduling injection molding operations with multiple resource constraints and sequence dependent setup times and costs. Comput. Oper. Res. **32**(11), 2987–3005 (2005). https://doi.org/10.1016/j.cor.2004.04.012. http://www.sciencedirect.com/science/article/pii/S0305054804000899

8. Dorigo, M., Maniezzo, V., Colorni, A.: The ant system: optimization by a colony or cooperating agents. IEEE Trans. Syst. Man Cybern.-Part B **26**, 29–41 (1996)

9. Dorigo, M., Stützle, T.: Ant Colony Optimization. MIT Press, Cambridge (2004)

10. Dorigo, M., Blum, C.: Ant colony optimization theory: a survey. Theor. Comput. Sci. **344**(2–3), 243–278 (2005). https://doi.org/10.1016/j.tcs.2005.05.020

11. Dorigo, M., Gambardella, L.M.: Ant colony system: a cooperative learning approach to the traveling salesman problem. IEEE Trans. Evol. Comput. **1**, 53–66 (1997)

12. Ham, A.: Scheduling of dual resource constrained lithography production: using CP and MIP/CP. IEEE Trans. Semicond. Manuf. **31**(1), 52–61 (2018). https://doi.org/10.1109/TSM.2017.2768899

13. Li, J., Huang, Y., Niu, X.: A branch population genetic algorithm for dual-resource constrained job shop scheduling problem. Comput. Ind. Eng. **102**, 113–131 (2016). https://doi.org/10.1016/j.cie.2016.10.012. http://www.sciencedirect.com/science/article/pii/S0360835216303813

14. Liao, C.J., Liao, L.M.: Improved MILP models for two-machine flowshop with batch processing machines. Math. Comput. Model. **48**(7), 1254–1264 (2008). https://doi.org/10.1016/j.mcm.2008.01.001. http://www.sciencedirect.com/science/article/pii/S089571770800040X

15. Lpez-Ibez, M., Dubois-Lacoste, J., Cceres, L.P., Birattari, M., Sttzle, T.: The irace package: iterated racing for automatic algorithm configuration. Oper. Res. Perspect. **3**, 43–58 (2016)

16. Merkle, D., Middendorf, M., Schmeck, H.: Ant colony optimization for resource-constrained project scheduling. IEEE Trans. Evol. Comput. **6**(4), 333–346 (2002). https://doi.org/10.1109/TEVC.2002.802450

17. Rajendran, C., Ziegler, H.: Ant-colony algorithms for permutation flowshop scheduling to minimize makespan/total flowtime of jobs. Eur. J. Oper. Res. **155**(2), 426–438 (2004). http://ideas.repec.org/a/eee/ejores/v155y2004i2p426-438.html

18. Smith, W.E.: Various optimizers for single state production. Naval Res. Logist. Q. **3**, 59–66 (1956)

19. Tang, L., Liu, P.: Minimizing makespan in a two-machine flowshop scheduling with batching and release time. Math. Comput. Model. **49**(5), 1071–1077 (2009). https://doi.org/10.1016/j.mcm.2008.09.012. http://www.sciencedirect.com/science/article/pii/S0895717708003476

20. Wang, W., Ma, C., Bao, Z., Ren, X.: A multi-objective model of integrated collaborative planning and scheduling for dual-resource and its algorithm. In: 2016 8th International Conference on Intelligent Human-Machine Systems and Cybernetics (IHMSC), vol. 1, pp. 400–403, August 2016. https://doi.org/10.1109/IHMSC.2016.154

21. Xu, J., Xu, X., Xie, S.: Recent developments in dual resource constrained (DRC) system research. Eur. J. Oper. Res. **215**(2), 309–318 (2011). https://doi.org/10.1016/j.ejor.2011.03.004. http://www.sciencedirect.com/science/article/pii/S0377221711002153

22. Zaerpour, F., Bischak, D.P., Menezes, M.B.: Coordinated lab-clinics: a tactical assignment problem in healthcare. Eur. J. Oper. Res. **263**(1), 283–294 (2017). https://doi.org/10.1016/j.ejor.2017.05.012. http://www.sciencedirect.com/science/article/pii/S037722171730440X

23. Zhang, Y., Liu, S., Sun, S.: Clustering and genetic algorithm based hybrid flowshop scheduling with multiple operations. Math. Prob. Eng. **2014**, 8 p. (2014). Article ID 167073. https://doi.org/10.1155/2014/167073

Optimization of the Velocity Profile of a Solar Car Used in the Atacama Desert

Dagoberto Cifuentes$^{(\boxtimes)}$ and Lorena Pradenas

University of Concepcion, Edmundo Larenas 219, Concepcion, Chile
{dagocifuentes,lpradena}@udec.cl
http://docindus.udec.cl/

Abstract. Global energy demand has undergone a substantial increase in past decades because of the rapid increase of the global population and the energetic consumption of new production technologies. As a result, a change is necessary in the global energy generating matrix, in which the sources originate primarily from renewable energy sources. The main renewable energy source may be solar energy, and one of its applications is solar mobility. A world-class solar racing car exists that requires a rational use of velocity and energy to minimize the time spent in a race. A total of three search metaheuristics were tested to achieve an efficient velocity profile for this car in the Atacama 2018 Solar Race: Genetic Algorithm, Simulated Annealing and Iterated Local Search. The three methods provided similar results, with Simulated Annealing being the one that provided better solutions.

Keywords: Hybrid electric vehicle · Energy management
Metaheuristics · Solar competition

1 Introduction

World's energy demand has significantly risen during the last decades due to the growth of global population and the increasing consumption of newer production technologies. As a result, the amount of energy generation has grown at the same rate to meet the global requirements. Therefore, it is necessary to consider the use of renewable energy sources, such as solar energy, wind energy, hydropower and geothermal. Out of these renewable sources, solar energy may be the best option because of the following reasons: (i) it is the most abundant source of renewable of energy in the Earth, receiving approximately 1.8×10^{14} [kW], fully covering the global energy demand; (ii) it is a virtually inexhaustible source of energy; (iii) its utilization causes no environmental damage whatsoever; (iv) it may be used at several scales and at both industrial and domestic levels [1]. Solar energy has various forms of use, one of these being its use in solar vehicles. The amount of solar energy collected by the exposed surface of the domestic vehicles in some parts of the world is able to propel them for a distance of 30 to 50 Km

© Springer Nature Switzerland AG 2019
M. J. Blesa Aguilera et al. (Eds.): HM 2019, LNCS 11299, pp. 164–171, 2019.
https://doi.org/10.1007/978-3-030-05983-5_12

per day, which is sufficient for a city car. Therefore, its domestic use is possible without the need to charge its battery in the main power grid [2].

Atacama Desert, located in Chile, receives a solar irradiance level above 7 kWh/(m2d), which is one of the largest in the world. The central region of the country receives a solar irradiance level above 5 kWh/(m2d), thus making Chile a prime candidate for the use of solar energy [3]. The increased solar irradiance levels in the Atacama Desert occur because of its location in the "solar belt", a region of the planet constrained between latitudes 40°N and 40°S. Furthermore, because of its altitude, the thickness of the atmosphere is reduced [4].

DAS-UdeC team built the solar vehicle AntuNekul II for solar racing vehicles competitions. There have been attempts to generate an efficient policy for the management of the speed and battery of the vehicle, although it has not been conducted because of their unsatisfactory results [5]. It is the pilot's task to perform the adjustment of the vehicle's velocity at all times, which leads to sub-optimal results for the racing times.

In this study, we propose a formal method of velocity and energy management for AntuNekul II. This car is a state-of-the-art solar competition car, but lacks a method to efficiently manage its velocity and energy system. Three metaheuristic algorithms were programmed in C++11: Simulated Annealing (SA), Iterated Local Search (ILS) and Genetic Algorithm (GA) [6]. Three test instances are used, and then, results and efficiency of each algorithm are evaluated.

2   Methodology

The objective of the study is to obtain an optimal velocity profile for AntuNekul II, to be used in the 2018 Atacama Solar Challenge, subject to physical and regulatory constraints.

2.1   Description of AntuNekul II

AntuNekul II is a hybrid electric vehicle (HEV). The power system of AntuNekul II is composed by an MPPT, photovoltaic cells, electric motor, controller and battery [5]. It has two power sources: photovoltaic cells that capture the incident solar radiation and the battery that transforms the stored chemical energy into electric power. The MPPT allows for optimization of the power provided by the solar panel through an adjustment of voltage and current supplied to the controller. The controller is responsible to determine the different operating modes, controlling the delivered current, velocity and acceleration of the HEV.

2.2   Problem Characterization

The aim of the optimization is to obtain an optimal velocity profile for the 2018 Atacama Solar race that minimizes the time spent on the route. The total length of the race is divided into segments. A total of 47 reference points were identified along the route, thus obtaining 46 segments between these reference points, as shown in Fig. 1. The segments do not necessarily have the same length or inclination. Each reference point has an hour h_i and a state of charge of the battery (SOC), B_i, associated. Each segment has a velocity v_i, a received solar energy $E_{s,i}$ and a demanded energy $E_{d,i}$ associated. The value of B_i depends on B_{i-1}, $E_{s,i}$ and $E_{d,i}$ [7].

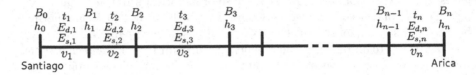

Fig. 1. Segmentation of the route from Santiago to Arica.

2.3   Solar Irradiance Estimation

As described by [5], a second-degree polynomial curve can be adjusted to mean solar irradiance hourly measurements [8], obtaining the total incident energy per $[m^2]$ in any segment of the race. This outcome is achieved through the use of the definite integral of the function between two hours h_{i-1} and h_i, as shown in Eq. 1.

$$E_{s,i}[\frac{J}{m^2}] = D + 3600 \times (A \times [h_{i-1}^3 - h_i^3] +$$
$$B \times [h_i^2 - h_{i-1}^2] + \tag{1}$$
$$C \times [h_{i-1} - h_i])$$

Parameters A, B and C take different values for each city, as presented in Table 1. These parameters are valid for all segments according to the reference point they represent, until reaching the reference point that corresponds to the next city (i.e. If the closest city to segment i is Santiago, the value for the parameters A, B, C and D for segment i, are taken from the first column of Table 1). Parameter D has a value equal to 0, except when there is a change of day, i.e., the current time is less than the previous time ($h_i < h_{i-1}$). This parameter is equivalent to the radiation obtained during a whole day per m^2. Once $E_{s,i}$ is obtained, the energy received by the HEV is obtained from Eq. 2. Parameters A_{panels}, Ef_{MPPT} and Ef_{panels} have values of 6 $[m^2]$, 0.99 and 0.226, respectively, corresponding to the area of the solar panels, efficiency of the MPPT and the efficiency of the solar panels [5].

$$E_{s,i,HEV}[J] = E_{s,i}[\frac{J}{m^2}] \times A_{panels}[m^2] \times Ef_{MPPT} \times Ef_{panels} \tag{2}$$

Table 1. Parameters A, B, C y D for Eq. 1

	Santiago	Valparaiso	La Serena	Copiapo	Antofagasta	Calama
A	6.75	7.07	7.31	9.70	9.29	11.74
B	243.07	254.67	263.26	349.32	334.48	422.52
C	2328.90	2440.80	2529.40	3363.20	3329.40	4082.10
D	19404101	20301926	20707170	27184193	25630889	32253541

2.4 Dynamic Model of the HEV

To calculate the demanded energy on each segment of the race by the vehicle, a dynamic model of the HEV is constructed. Figure 2 shows the free body diagram of a car while ascending a slope with an inclination angle of θ [5]. Only the x-axis is considered because there is no displacement in the y-axis. The following incident forces on the vehicle are considered: force exerted by the engine, friction force by the rolling resistance, aerodynamic drag, and the component in the x-axis of the gravity force. [5] shows that Eq. 3 describes the consumed energy $E_{c,i}[J]$ by the vehicle in each segment, where C_R corresponds to the friction coefficient and has a value of 0.911, the vehicle mass m of has a value of 299.5 [kg], constant g has a value of 9.8 [m/s²], λ has a value of 0.1125. θ_i represents the inclination of segment i. t_i and V_i represent time spent and velocity of the vehicle on segment i respectively.

$$E_{c,i}[J] = t_i(C_R mg \cos \theta_i V_i + \lambda V_i^3 + mg \sin \theta_i V_i) \tag{3}$$

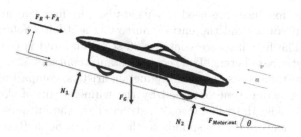

Fig. 2. Free body diagram of AntuNekul II.

2.5 Velocity Profile Optimization

Let $\mathbf{V} = [v_1, v_2, \ldots, v_N]$ be a vector that contains a velocity profile for the N segments in which the race is divided. Let $\mathbf{L} = [l_1, l_2, \ldots, l_N]$ be a vector that contains the lengths l_i of each segment of the race. Likewise, let $\mathbf{T} = [t_1, t_2, \ldots, t_N]$ be a vector that contains the times t_i used to travel each segment of the race, where t_i is the quotient of l_i and v_i. The solution of the problem consists in minimizing objective function 4, or fitness function regarding Genetic Algorithm. To find an efficient velocity profile, the following search techniques are used: Iterated Local Search, Simulated Annealing and Genetic Algorithm.

$$min F_{obj} = \sum_{i=1}^{N} t_i \tag{4}$$

2.6   Constraints

Restrictions for this problem can be classified according to their nature [5]. These restrictions may be regulatory, referring to the rules of the race [9], or physical, referring to the HEV itself. The physical constraint is that the SOC of the battery cannot be less than 5%. Because the battery of AntuNekul II has a capacity of 17496 [kJ], the minimum SOC corresponds to 874.8 [kJ]. Regulatory constraints are:

1. The race starts at 08:30 on the first day and at 08:00 the following days. The race stops at 17:00 everyday.
2. The HEV is not allowed to recharge its battery in the night stop, that is, between 17:00 of one day and 08:00 of the next day.
3. The HEV will be disqualified if it travels at a speed less than 50 [km/h].
4. The maximum speed can be of 100 [km/h], according to Chilean regulations.
5. There is a 30-min solar stop every day of the race.

2.7   Test Instances

A total of three instances are used in this study. The first two instances correspond to a half course and the entire course of the *World Solar Challenge* in Australia [7]. The first instance contains 26 segments, and is used to calibrate the metaheuristic parameters. The second instance contains 52 segments and is used to validate the results obtained by metaheuristics, comparing the obtained results with the actual time obtained by the winner team of the World Solar Challenge 2017. The third instance, constructed by the authors of this study, contains 46 segments and corresponds to the Atacama Solar Race 2018.

2.8   Metaheuristics

As described before, three search metaheuristics were used in this study: SA, ILS and GA. The solution representation consists of a vector **V** of length n, containing the vehicle's constant velocity v_i for each of the n segments of the race, as described in Sect. 2.5. SA and ILS generate a random feasible initial solution, GA generates a random feasible initial population. To assure the feasibility of a solution, the algorithm assigns a uniformly distributed random number between 50 and 100 to each segment. Later, it calculates the received and demanded energy for each segment, thus obtaining the SOC for each reference point. If it violates the physical constraint described in Sect. 2.6, a new solution is generated.

Local Search on ILS is performed using a 2-opt approach [10], while Perturbation is performed by assigning a new random velocity to two random segments. For SA, the correct definition of the neighborhood of the solutions has a notorious impact on the performance of the algorithm. If the neighborhood is too small, the algorithm cannot explore enough solutions before convergence. If it is too large, the algorithm performs practically a random search [11]. A neighbor is defined as a vector containing exactly two different components. On GA, crossover is performed using 2-opt, and mutation is performed by assigning a random new velocity to a random chromosome.

3   Results

The three search metaheuristics: ILS, SA and GA were programmed in C++ in the IDE Code::Blocks Release 16.01 rev 10702. To evaluate the quality of the algorithms, two parameters are measured: total time of the race and execution time of the algorithm. The total time of the race provides a direct measure of the quality of the solution because this is the value to be minimized. The execution time of the algorithm provides a measure of its efficiency.

3.1   World Solar Challenge Instance

To analyze the metaheuristics, a comparison of the results with a known efficient solution is required. Because there is no benchmark solution available, instance 2 is used, corresponding to the entire World Solar Challenge length. Thus, the results obtained by the algorithms for this instance are compared with the actual time obtained by the winning team of the 2017 edition. The winning team arrived on the seventh day of the race, at 14:10. Note that the World Solar Challenge has the same rules than the Atacama Solar Race, except that the area of the solar panels must be of a maximum of $4\,[\text{m}^2]$ instead of $6\,[\text{m}^2]$. For this reason, this parameter must be adjusted in the three algorithms for this instance. It is assumed that the incident radiation is similar to the Chilean case because Australia is in the same hemisphere as Chile, and the race was held in the same month as the Atacama Solar race. A total of 5 replicas are executed for each of the algorithms. Table 2 shows the obtained results. The first row corresponds to the best value, the second row corresponds to the average value and the last row contains the worst values. The three methods are found to provide better results than those of the actual winning team. Therefore, it can be concluded that the three methods provide good quality results.

3.2   Atacama Solar Race 2018 Instance

The results obtained by the three algorithms for instance 3, corresponding to Atacama Solar Race 2018, are shown. Five replicas are performed for each algorithm. The results obtained for instance 3 are shown in Table 3. The first row

Table 2. Results for the example of the World Solar Challenge

ILS			SA			GA		
R [s]	ET [s]	Hour	R [s]	ET [s]	Hour	R [s]	ET [s]	Hour
193844	2.33	11:02	193900	10.25	11:02	195117	18.50	11:04
194125	4.87	11:03	193994	11.40	11:02	195339	21.36	11:05
194312	7.47	11:03	194084	11.68	11:02	195706	25.04	11:06

corresponds to the best value, the second row corresponds to the average value and the last row contains the worst values. It is shown that Simulated Annealing delivers the best results, and Iterated Local search delivers the shortest execution times. Genetic Algorithm provides the worst results, in addition to being the slowest algorithm.

Table 3. Results obtained for the three methods for the Atacama Solar Race 2018.

ILS		SA		GA	
R [s]	ET [s]	R [s]	ET [s]	R [s]	ET [s]
146027	1.11	145710	11.17	146865	14.85
146345	1.91	145792	11.88	147735	17.22
146950	2.37	145904	12.63	148118	20.63

4   Discussion and Conclusions

A total of three metaheuristics were designed, calibrated and tested, namely, ILS, SA and GA, to obtain an efficient velocity profile, with the goal of minimizing the total time spent on the race. These algorithms were developed for any instance described by reference points that contain two parameters: distance from the start of the race and altitude. The three algorithms function in a similar manner: ILS and GA generate a feasible initial solution, whereas GA generates a population of initial feasible solutions. These algorithms improve this solution through an iterative search process in the space of solutions, whereas GA generates a population of initial feasible solutions, and in each iteration, the population is improved, subject to the described fitness function.

A total of three test instances in the problem were used: the first two correspond to the World Solar Challenge, and the last one corresponds to the Atacama Solar Race 2018. The latter was created by the author, with the assistance of Google Maps and Google Earth software. Instances 1 and 2 were used to calibrate the parameters of the algorithms and to validate the results obtained, respectively. The third instance was used to create an efficient velocity profile for the Atacama Solar Race 2018.

Section 3 shows that the three algorithms had a shorter driving time than the winning team of the World Solar Challenge 2017; thus, it is concluded that reliable and good quality results are provided. Both in instances 2 and 3 the algorithm SA is the one that provides the best results, ILS provides a shorter execution time, whereas GA provides the worst results in a longer execution time.

References

1. Kannan, N., Vakeesan, D.: Solar energy for future world: - a review. Renew. Sustain. Energy Rev. **62**, 1092–1105 (2016)
2. Solar cars? Not just the realm of comic books any more. Electr. J. **30**(10), 83–84 (2017)
3. Haas, J., et al.: Sunset or sunrise? Understanding the barriers and options for the massive deployment of solar technologies in Chile. Energy Policy **112**, 399–414 (2018)
4. Cornejo, L., Martín-Pomares, L., Alarcon, D., Blanco, J., Polo, J.: A through analysis of solar irradiation measurements in the region of Arica Parinacota, Chile. Renew. Energy **112**, 197–208 (2017)
5. Cifuentes, D.: Optimizing the performance of a competition hybridelectric vehicle. Master's thesis, University of Concepcion (2018)
6. Talbi, E.-G.: Metaheuristics: From Design to Implementation. Wiley Publishing, Hoboken (2009)
7. Elshafei, M., Al-Qutub, A., Saif, A.W.A.: Solar car optimization for the world solar challenge. In: 2016 13th International Multi-Conference on Systems, Signals Devices (SSD), pp. 751–756, March 2016
8. P. UTFSM, CNE: Solar irradiance over territories of the Republic of Chile. Registro solarimétrico (2008). http://www.plataformacaldera.cl/biblioteca/589/w3-article-64683.html. Accessed 10 Mar 2018
9. LA RUTA SOLAR Project: Atacama Solar Challenge 2018 Rulebook. Solar Vehicles Category (2017). http://www.carrerasolar.com/wp-content/uploads/2017/09/Bases-Solares-Carrera-Solar-Atacama-v1.0.pdf. Accessed 10 Mar 2018
10. Croes, G.A.: A method for solving traveling-salesman problems. Oper. Res. **6**(6), 791–812 (1958)
11. Goldstein, L., Waterman, M.: Neighborhood size in the simulated annealing algorithm. Am. J. Math. Manag. Sci. **8**(3–4), 409–423 (1988)

Local Search Methods
for the MRCPSP-Energy

André Renato Villela da Silva[1(✉)] and Luiz Satoru Ochi[2]

[1] Institute of Science and Technology, Federal Fluminense University,
Rio das Ostras, Brazil
arvsilva@id.uff.br
[2] Institute of Computing, Federal Fluminense University, Niterói, Brazil
satoru@ic.uff.br

Abstract. The Multi-Mode Resource-Constrained Project Scheduling
Problem with energy saving (MRCPSP-energy) is a variant of the classical Resource-Constrained Project Scheduling Problem (RCPSP). In this
variant, the execution of each job must take into account the job duration and the energy spent to execute that job, which are conflicting. The
objective is to minimize both makespan and total energy consumption.
This work proposes two local search methods to improve a large dataset
of inputs. One of them is a restricted version of a Mixed-Integer Programming formulation and the other one is a heuristic local search called H.
The computational experiments showed that the hybrid method with
the H algorithm obtained better solutions and is competitive with the
literature results.

Keywords: MRCPSP · MRCPSP-energy · Heuristics
Local search procedures · Hybrid heuristics

1 Introduction

The Project Scheduling Problem (PSP) is one of the most studied subjects in
the combinatorial optimization field due to its application in different areas.
A classical variant is the Multi-Mode Resource-Constrained Project Scheduling
Problem (MRCPSP), where the jobs can be executed in different ways (modes).
Each mode assumes a possible distinct job duration and resources needed to execute the job. The objective is to minimize the makespan, i.e. the finish time of the
last job, without violations in the job precedences and the resources availability.

Nowadays, there is great concern in reducing the consumption of energy in
manufacturing, without negatively impacting final production. The MRCPSP
with energy saving (MRCPSP-energy) was proposed in [14] to model such situations. In the MRCPSP-energy, the job modes require the same quantity of
available resources, but have a duration and an amount of energy needed that
may vary from mode to mode. The objective is to minimize both the makespan

© Springer Nature Switzerland AG 2019
M. J. Blesa Aguilera et al. (Eds.): HM 2019, LNCS 11299, pp. 172–179, 2019.
https://doi.org/10.1007/978-3-030-05983-5_13

and the total energy spent. However, these two elements are conflicting, since a lower energy consumption implies in a longer job duration.

Instead of employing a bi-objective function, the authors adopted a new metric, called "relative project efficiency" (η). Suppose a solution S has makespan S_m and total energy consumption S_e. The project relative efficiency $\eta = \frac{LB0}{S_m} * \frac{E_{min}}{S_e}$, where $LB0$ is the makespan lower bound given by the Critical Path Method and E_{min} is the sum of the minimum energy mode of all jobs. So, the higher the efficiency the better the solution is. The MRCPSP-energy has many applications in industry and manufacturing where the modern machinery is capable of operating in different energy-saving profiles.

This work presents two local search approaches to improve basic solutions for the MRCPSP-energy. The first one is restricted Mixed-Integer Programming formulation, where additional constraints are heuristically generated. The second approach is a local search procedure which adapts its movements according to previous improvements obtained. Both methods have not yet been proposed for the MRCPSP-energy.

The remainder of this paper is given as follows. The Sect. 2 highlights the still scarce literature about the problem. The Sects. 3 and 4 present some proposed local search methods and the computational results, respectively. The Sect. 5 brings out the conclusions of this work.

2   Literature Review

The literature about RCPSP and its variants is so extensive that there are some surveys that seek to classify or give a general notation for problem models as presented in [1,4,7]. Almost every optimization combinatorial method has been used from exact approaches [5,18], through classical heuristic algorithms [2,3,6] to hybrid methods [10,16,17].

In [14], the MRCPSP-energy and a benchmark set of instances (2040 artificial projects with 30–120 jobs) were proposed. In [15], the first heuristic (Ant Colony Optimization - ACO) and Mixed-Integer Programming (MIP) formulation were proposed. Both performed relatively well, finding several optimal solutions in small instances, but they struggled on large instances, where no good bounds were provided.

In [12], the authors proposed a Genetic Algorithm (GA) based on a double list codification. These two lists are called Activity List (a priority list used to schedule the jobs of the project) and Mode List (a n-dimensional array, where each element indicates the execution mode of a job). The authors proposed exclusive operators for each list and used them in different phases of the GA. In the same paper, a Mixed Integer Non-Linear Programming formulation is also proposed (the objective function is non-linear). This formulation is quite similar to that in [15], although the linearization of the objective function is not presented. The GA achieved almost the same results of the mathematical formulation taking a much smaller computational time. Comparing [12] and [15], the MIP formulation of the first work had a better average efficiency while the GA slightly outperformed the ACO.

3   Proposed Local Search Approaches

Formally, [12] described the problem as a project that consists of a set of n activities (jobs) $I = 0, ..., i, ..., n$, a set B of K^ρ shared renewable resources $B = 1, ..., b, ..., K^\rho$, and an available amount R_b^ρ of every renewable resource. Each i has M_i execution modes, where each mode $m \in M_i$ requires a nonpreemptive execution time d_{im}, a total of r_{ib}^ρ renewable resources of type b, and an amount of energy e_{im} for its realization. The activities must respect precedence among them and the objective is to minimize the relative project efficiency (η), which comprises both project makespan and the total energy consumption.

To create a base of solutions on which local search approaches can be run and analyzed, several instances from [14] have been selected at random. The Ant Colony Optimization (ACO) algorithm [15] was used to produce these solutions for each selected instance. Each time a better solution was found, it was inserted into an elite set of solutions that will be used for local search approaches with the MIP formulation. The final solution for this instance is the best solution found during the execution of the algorithm.

3.1   MIP Extra Restrictions

Combinatorial optimization softwares such as Cplex (www.ibm.com/analytics/cplex-optimizer) or Gurobi (www.gurobi.com) have been used in recent years as a tool to perform local search approaches by reducing the number of variables or increasing the number of constraints. In this case, it is expected that a more limited solution space will favor the faster computation of a better quality solution.

This way, we seek to use a more restricted version of the mathematical formulation present in [15]. Each elite set of the dataset is then analyzed in 5 characteristics: (1) The interval between the largest and the smallest makespan; (2) the interval between the highest and lowest total energy consumption; (3) both of the previous intervals; (4) the interval between the highest and lowest start times of each job; (5) the interval between the highest and lowest finish time of each job.

For each characteristic analyzed, additional constraints will be created and inserted into the original mathematical formulation, so that a feasible solution must also respect them. These limitations will use the intervals described above, applying a small Δ value to the bounds of the intervals.

For example, suppose that the solutions of the elite set have makespan between 20 and 30. If $\Delta = \pm 10\%$, the mathematical formulation will require a feasible solution to have makespan between 18 ($= 20 \times 0.9$) and 33 ($= 30 \times 1.1$). If the limitation is given by the energy consumption, a feasible solution must have similar lower and upper bounds in this characteristic. Regarding the jobs execution modes, only those modes that are used in at least one elite set solution will be allowed by the restricted formulation. The limitations given by the start or finish intervals of each job will use a δ value that will be added to or subtracted from the bounds of those intervals.

We adopted a set of elite solutions and their respective intervals because applying extra constraints on a single input (solution) seldom allowed the finding of better solutions.

3.2 Heuristic Approach

We also propose a new heuristic approach, called H, for the MRCPSP-energy, in order to be used as a local search procedure. The algorithm H takes a sequence of pairs (i, m), where i indicates a job and m indicates an execution mode. This sequence is retrieved from an initial feasible solution given to the algorithm.

The project jobs are scheduled according to that sequence, at the earliest possible time. This time depends on the resource availability and the predecessors finishing time. In that sequence, if job i is predecessor of job j, i must appear before j.

In order to explore the solution neighborhood, some operations are proposed. The first one changes the execution mode of a randomly selected job to a new mode also randomly chosen. The second and the third operations, respectively, postpone or anticipate a randomly selected job as much as possible, without violating the jobs precedence order. The last operation swaps a job with its successor in the sequence, if it is possible. Every operation can be repeatedly executed a given number of times before the scheduling algorithm is called to compute the solution value.

The pseudo-code can be seen in Algorithm 1. The scheduling algorithm is called inside the $Execute(..)$ procedure (l. 14) to evaluate the modified solution.

Algorithm 1. H(input: feasible solution S_0)

1: $S^* \leftarrow S \leftarrow S_0$
2: $iter \leftarrow 0$
3: $last \leftarrow 0$
4: $factor \leftarrow 1.0$
5: **while** stop criterion is not met **do**
6:     **if** $iter - last \geq Limit1$ **then**
7:         $factor \leftarrow Slack$
8:     **else if** $iter - last \geq Limit2$ **then**
9:         $S \leftarrow S_0$
10:        $last \leftarrow iter$
11:    **end if**
12:    $op \leftarrow ChooseOperation()$
13:    $x \leftarrow ChooseQuantity()$
14:    $S' \leftarrow Execute(S, op, x)$
15:    **if** $value(S') < value(S) * factor$ **then**
16:        $S \leftarrow S'$
17:        $last \leftarrow iter$
18:        $factor \leftarrow 1.0$
19:        **if** $value(S') < value(S^*)$ **then**
20:            $S^* \leftarrow S'$
21:            $UpdateProbabilities()$
22:        **end if**
23:    **end if**
24: **end while**
25: **return** S^*

At the beginning, the algorithm uses an initial feasible solution (l. 1). The current iteration and the last improvement counters are initialized (ls. 2–3) as well as the acceptance factor (l. 4). If the last improvement occurred many iterations ago (l. 6), a first diversification mechanism is applied, allowing new solutions to be a little worse than the incumbent solution S (l. 15). The second diversification is triggered if H executes further without improvements. In this case, H restarts the incumbent solution (ls. 8–11).

At that point, the algorithm performs some modifications to the solution S, choosing an operation and a number of repetitions (ls. 12–14). It is important to notice that the scheduling (evaluation) procedure is executed once per iteration, only after all repetitions for the chosen operation is performed. Thus, these repetitions modifies the current solution "blindly". This scheme is used to avoid the hill-climbing local optima convergence.

If the modified solution is acceptable (l. 15), it replaces the incumbent solution S (ls. 16–18). If S' is the best solution found so far, the operations and repetitions probabilities are updated. This update procedure is the H intensification mechanism. Initially, all options have the same probability. Every time a new best solution is found, the operation selected at that iteration and the number of repetitions made receive a bonus and become more likely of being chosen afterward.

The selections at lines 12 and 13 use the roulette technique. So, a simple way of implementing the whole method consists in starting each option probability with an integer p and the given bonus increases the specific probability by an amount k. This strategy was used in [9,13], for example. In our experiments, the following parameter values were used: $p = 10$, $k = 1$, $Limit1 = 1000$, $Limit2 = 5000$ and $Slack = 1.1$. Each operation may be performed up to 5 times at each iteration. All parameters were defined after several preliminary tests, combining many different values. In these tests, $Limit2 = 10000$ achieved statistically similar results.

4   Computational Results

The instances used in the computational experiments were retrieved from [11] and the computational environment is similar to [15] and to [12]: Intel I7-3630QM, 8 GB RAM, Windows 8.0, Cplex 12.7.3. All algorithms were implemented using C++.

Each solution of the dataset was generated by the ACO algorithm with stopping criterion of 24000 schedules, which will be explained later. This amount of schedules corresponds to 80 rounds of 300 ants each. The average computational time was 0.8, 1.4, 2.1 and 2.9 s for instances with 30, 60, 90 and 120 jobs, respectively. The average efficiency can be seen in Table 2.

In the first experiment, the five characteristics extracted from the elite set were tested by inserting their additional constraints to the original mathematical formulation. As mentioned previously, each characteristic was related to some Δ values. Local search approaches had a time limit of 5, 10, 15 and 20 min for instances with 30, 60, 90 and 120 jobs, respectively. The Table 1 shows the five largest average efficiency increase of the best solution obtained by the local search in relation to the original solution (generated by the ACO).

Table 1. Best average improvement obtained by the local search implied by each analyzed characteristic. (1) The interval between the largest and the smallest makespan; (3) both makespan and total energy consumption intervals; (4) the interval between the highest and lowest start times of each job.

Charac.	Δ	Improv.
4	±3	0.0255
4	±2	0.0255
4	±1	0.0252
1	±15%	0.0248
3	±15%	0.0248

The results indicated that the largest increments are produced by performing the local search using the makespan alone (characteristic 1) or in conjunction with the total energy consumed (characteristic 3), or by limiting the start times of the tasks (characteristic 4). In absolute terms, the total energy consumed varied between 3 and 8 times the projects makespan. Thus, the reduction of only one unit in makespan represents a large increase in the solution efficiency. The exploration of characteristics 1 and 3 privileges the reduction of makespan. In relation to characteristic 4 (interval between the largest and smallest jobs start times), the results showed that the larger the Δ value, the larger the space of solutions to be explored by the local search. So, the chance of finding a better solution also increases.

The Table 2 (third column) presents the improvements obtained by the local search approaches using the characteristics analyzed. The average computational time was 14, 272, 631 and 941 s for instances with 30, 60, 90 and 120 jobs, respectively. Approximately 38% of the local search executions finished before the time limit.

In the last computational experiment, the solutions generated by the ACO are given to the algorithm H, which executes 26000 iterations for each instance. In this way, the joint execution of the ACO+H represents a total of 50000 schedules. This stopping criterion is very common in the literature and it has been suggested by [8] and followed by [12]. Since the algorithm H is stochastic, 20 runs are performed for each input. Table 2 (fifth column) shows the average improvements obtained for each set of instances.

The fourth and sixth columns of Table 2 shows that the solutions produced by the algorithm H have a higher average efficiency value than the solutions produced by the restricted mathematical formulation. In addition, the computational time of the algorithm H is much lower, being approximately 0.5, 0.9, 1.7 and 3.1 s, respectively, for instances with 30, 60, 90 and 120 jobs. The best solutions for each execution are usually found in less than half that time.

Table 2. Average improvement and final results for each instance size.

Jobs	ACO	L.S. improv.	L.S. final	ACO+H improv.	ACO+H final	[15]	[12]
30	0.6411	0.0105	0.6516	0.0165	0.6576	0.6495	0.6564
60	0.6610	0.0367	0.6977	0.0407	0.7017	0.6666	0.6919
90	0.6630	0.0355	0.6985	0.0466	0.7096	0.6757	Not avail.
120	0.5098	0.0467	0.5565	0.0629	0.5725	0.5182	0.5590

Finally, the Table 2 also compares the best solutions produced by the hybrid ACO+H algorithm with the results provided in [12] and [15] (the last two columns). The results of the ACO+H hybrid method are slightly higher than those reported in the literature. However, it is not possible to state that one method is better than the other because in this work, the generation of solutions for the dataset used a subset of randomly chosen instances in order to show the potential for improvement of the proposed local search methods.

5   Concluding Remarks

The MRCPSP-energy is a new variant of the classical RCPSP where the jobs can be executed with multiple modes, representing distinct job durations and a required amount of energy to execute that job. The objective is to minimize the relative project efficiency (η), which comprises the project makespan and the total energy consumption.

As the MRCPSP-energy was recently proposed, the literature presents only a couple of approaches. Two local search methods are proposed in this work. The first one consists in inserting additional constraints into a MIP formulation in order to restrict the local search to look for a better solution according to some characteristics retrieved from a elite set of instances. The second method consists in applying a local search heuristic (called H) to improve the solutions provided by another heuristic algorithm of the literature.

After several experiments and comparisons, the hybrid method ACO+H obtained better results for the compiled dataset. The hybrid algorithm is also competitive with the literature results.

Future works include more experiments with a larger set of instances to show how much robust the hybrid algorithm is. Another possibility consists in improving the literature mathematical formulations in order to produce more optimal solutions or, at least, to obtain improved bounds.

References

1. Blazewicz, J., Lenstra, J., Rinnooy Kan, A.: Scheduling subject to resource constraints: classification and complexity. Disc. App. Math. **5**, 11–24 (1983)
2. Boctor, F.F.: Some efficient multi-heuristic procedures for resource constrained project scheduling. Eur. J. Oper. Res. **49**, 3–13 (1990)

3. Bouleimen, K., Lecocq, H.: A new efficient simulated annealing algorithm for the resource-constrained project scheduling problem and its multiple mode version. Eur. J. Oper. Res. **149**, 268–281 (2003)
4. Brucker, P., Drexl, A., Möhring, R., Neumann, K., Pesch, E.: Resource-constrained project scheduling: notation, classification, models, and methods. Eur. J. Oper. Res. **112**, 3–41 (1999)
5. Demeulemeester, E., Herroelen, W.: A branch-and-bound procedure for multiple resource-constrained project scheduling problem. Manag. Sci. **38**, 1803–1818 (1992)
6. Gonçalves, J.F., Mendes, J.J.M., Resende, M.G.C.: A random key based genetic algorithm for the resource constrained project scheduling problems. Int. J. Prod. Res. **36**, 92–109 (2009)
7. Hartmann, S., Briskorn, D.: A survey of variants and extensions of the resource-constrained project scheduling problem. Eur. J. Oper. Res. **207**(1), 1–14 (2010)
8. Kolisch, R., Hartmann, S.: Experimental investigation of heuristics for resource-constrained project scheduling: an update. Eur. J. Oper. Res. **174**(1), 23–37 (2006)
9. Lu, Y., Benlic, U., Wu, Q.: A hybrid dynamic programming and memetic algorithm to the traveling salesman problem with hotel selection. Comput. Oper. Res. **90**, 193–207 (2018). https://doi.org/10.1016/j.cor.2017.09.008
10. Myszkowski, P.B., Skowroński, M.E., Olech, Ł.P., Oślizło, K.: Hybrid ant colony optimization in solving multi-skill resource-constrained project scheduling problem. Soft Comput. **19**(12), 3599–3619 (2015)
11. Morillo, D.: PSPLIB-ENERGY: a PSPLIB extension for evaluating energy optimization in MRCPSP. http://gps.webs.upv.es/psplib-energy/. Accessed 28 Nov 2017
12. Morillo, D., Barber, F., Salido, M.A.: Mode-based versus activity-based search for a nonredundant resolution of the multimode resource-constrained project scheduling problem. Mathem. Probl. Eng. **2017**, 15 (2017). https://doi.org/10.1155/2017/4627856. Article ID 4627856
13. Prais, M., Ribeiro, C.C.: Reactive GRASP: an application to a matrix decomposition problem in TDMA Traffic assignment. INFORMS J. Comput. **12**(3), 163–255 (2000)
14. Torres, D.M., Barber, F., Salido, M.A.: MRCPSP-ENERGY, un enfoque metaheurístico para problemas de programación de actividades basados en el uso de energía. In: Proceedings of XVIII Latin Ibero-American Conference on Operations Research (CLAIOXVIII), Santiago-Chile, October 2016
15. Villela da Silva, A.R.: Techniques to solve a multi-mode resource constrained project scheduling problem with energy saving. In: Proceedings of XIII Brazilian Congress on Computational Intelligence (XIII CBIC), Niterói-Brazil, October 2017
16. Tseng, L.-Y., Chen, S.-C.: A hybrid metaheuristic for the resource-constrained project scheduling problem. Eur. J. Oper. Res. **175**(2), 707–721 (2006)
17. Valls, V., Ballestin, F., Quintanilla, M.S.: A hybrid genetic algorithm for the resource-constrained project scheduling problem. Eur. J. Oper. Res. **185**(2), 495–508 (2008)
18. Zhu, G., Bard, J., Tu, G.: A branch-and-cut procedure for the multimode resource-constrained project-scheduling problem. J. Comput. **18**(3), 377–390 (2006)

Adaptation of Late Acceptance Hill Climbing Algorithm for Optimizing the Office-Space Allocation Problem

Asaju La'aro Bolaji[1]([✉]), Ikechi Michael[2], and Peter Bamidele Shola[3]

[1][*]Department of Computer Science, Faculty of Pure and Applied Sciences,
Federal University Wukari, Wukari, Taraba State, Nigeria
lbasaju@fuwukari.edu.ng
[2] Andela Office, 23 Olusoji Idowu Street, Ilupeju, Ikorodu Road, Lagos, Nigeria
mykehell123@gmail.com
[3] Department of Computer Science, University of Ilorin, Ilorin, Nigeria
shola.bp@unilorin.edu.ng

Abstract. Office-space-allocation (OFA) problem is a category of a timetabling problem that involves the distribution of a set of limited entities to a set of resources subject to satisfying a set of given constraints. The constraints in OFA problem is of two types: hard and soft. The hard constraints are the one that must be satisfied for the solution to be feasible while the violation of soft constraints is allowed but it must be reduced as much as possible. The quality of the OFA solution is determined by the satisfaction of the soft constraints in a feasible solution. The complexity of the OFA problem motivated the researchers in the domain of AI and Operational research to develop numerous metaheuristic-based techniques. Among recently introduced local search-based metaheuristic techniques that have been successfully utilized to solve complex optimization problem is the Late Acceptance Hill Climbing (LAHC) algorithm. This paper presents an adaptation of LAHC algorithm to tackle the OFA problem in which three neighbourhood structures are embedded with the operators of the LAHC algorithm in order to explore the solution space of the OFA efficiently. The benchmark instances proposed by the University of Nottingham and University of Wolverhampton datasets are employed in the evaluation of the proposed algorithm. The LAHC algorithm is able to produced one new result, two best results and competitive results when compared with the state-of-the-art methods.

Keywords: Timetabling problem · Office-space allocation
Metaheuristics · Late Acceptance Hill Climbing
Local search-based methods

1 Introduction

Many practical applications of the office-space allocation (OFA) problem to numerous organization have been proposed and tackled using several algorithmic

© Springer Nature Switzerland AG 2019
M. J. Blesa Aguilera et al. (Eds.): HM 2019, LNCS 11299, pp. 180–190, 2019.
https://doi.org/10.1007/978-3-030-05983-5_14

techniques by the workers in the field of timetabling and operational research over the last few decades. The OFA is a complex combinatorial optimization problem which is NP-hard in all its variation [1]. It involves allocating a set of resources limited to a set of spaces such that all resources are assigned to the required spaces subject to satisfying a set of constraints in order to achieve optimal utilization. Generally, the classification of constraints in OFA is grouped into hard and soft: the hard constraints in an OFA problem must be compulsorily satisfied for the solution to be feasible, whereas it is not mandatory to satisfy all the soft constraints but their satisfactions improved the quality of the solution. The quality of the solution to the OFA problem is measured by the satisfactions of number of soft constraints.

The algorithmic techniques employed by the researchers for the OFA are exact approaches, heuristics-based and metaheuristic-based approaches [2]. Few examples of the studies that proposed the exact methods for the OFA can be found in [3–5]. Numerous applications of metaheuristic-based algorithms for solving many complex optimization problems have been recorded over the past two decades. Metaheuristic-based algorithms that have been employed successfully in the field of timetabling to optimize OFA problem are classified into population-based and local search-based methods. Few examples of population-based methods proposed for the OFA include harmony search [6], genetic algorithm [7,8], and artificial bee colony algorithm [9]. These algorithms work with many solution and iteratively improve these solutions based on existing knowledge from previous searches. Similarly, local search-based approaches begin with one solution and iteratively improve this solution until a desired solution is achieved. Some examples of local search-based approaches employed for the OFA include, simulated annealing [7,10], tabu search [11] and hill climbing [7,11]. Furthermore, studies that proposed usage of hyper-heuristic and hybrid metaheuristic approaches for the OFA can be found in [12–14].

Late acceptance hill climbing (LAHC) algorithm is an extension of the classical hill climbing, recently introduced to tackled the examination timetabling problem by Burke and Bykov in 2008 [15]. Originally, the idea behind the LAHC is its late acceptance strategy that is utilized to prevent the algorithm from getting stuck in a local minimum that common to many greedy search algorithms. Due to its simplicity, it has been successfully adapted, applied, modified and hybridized by the researchers in the domain of artificial intelligence and operations research to tackle numerous optimization problems such as balancing two-sided assembly lines [16], course and examination timetabling [17], lock scheduling [18], liner shipping fleet repositioning [19], traveling salesman problem [20], patient admission scheduling problem [21]. The LAHC was also hybridized with other metaheuristics components in [22] and its applicability in the hyper-heuristics framework have been reported in [23,24]. However, none of the studies have reported it adaptation for tackling Office-space Allocation (OFA) problem. Thus, the main focus of this paper is to investigate the performance of LAHC algorithm for solving the OFA problem.

2 Office-Space Allocation Problem (OFA)

The formulation to the OFA involves a set of l entities, with dimensions $\eth = \{d_i | i = 0, \ldots, l-1\}$, and a set of m rooms with capacities $\mathfrak{c} = \{c_i | i = 0, \ldots, m-1\}$. The OFA solution is given by a two dimensional vector X of $[x_{i,j}]$ values, where $x_{i,j} = 1$ if the entity j (e_j) is assigned to room i (r_i). The main objective of the OFA problem is to generate a feasible solution with the best quality. The different constraints of the OFA problem considered in this study are:

- *No sharing* - this constraint specifies that the room of a particular entity should not be shared with another entity.
- *Be located in* - a specific room should be allocated to a particular entity.
- *Be adjacent to* - a particular entity should be assigned to room adjacent to another entity.
- *Be away* - room allocation of a particular entity should not close to another entity.
- *Be together with* - two particular entity should be assigned to the same room.
- *Be grouped with* - A group of entities should be assigned close to each other.

The solution to the OFA is evaluated based on the penalty cost $f(x)$ as specified in Eq. (1)

$$\min f(x) = f_1(x) + f_2(x). \tag{1}$$

subject to

$$\sum_{i=0}^{l-1} \sum_{j=0}^{m-1} x_{i,j} = 1. \tag{2}$$

where the space misuse function is given by $f_1(x)$ and violation of the soft constraints is computed by $f_2(x)$.

$$f_1(x) = \sum_{i=0}^{l-1} WP_i + \sum_{i=0}^{l-1} OP_i. \tag{3}$$

$$f_2(x) = \sum_{r=0}^{s-1} SCP_r. \tag{4}$$

where both WP_i and OP_i are the amount of space wasted or overused for each room i; SCP_r represents the penalty for violating the rth soft constraint. For each room only i one of WP_i or OP_i has a value greater than zero, and the amount of overused for each room i is computed as shown in Eqs. 5 and 6.

$$WP_i = \max(0, c_i - \sum_{j=0}^{m-1} x_{i,j} \cdot w_j). \tag{5}$$

$$OP_i = \max(0, 2(\sum_{j=0}^{m-1} x_{i,j} \cdot w_j - c_i)). \tag{6}$$

Where w_j is the space requirement of resource j and c_i represents the capacity of the room i.

3   Late Acceptance Hill Climbing Algorithm

Late Acceptance Hill Climbing (LAHC) algorithm is one of the newly introduced multi-purpose meta-heuristic techniques, proposed for examination timetabling problems by Burke and Bykov [15]. It belongs to the category of iterative search techniques which employs an advanced acceptance strategy in its operation. Typically, it is based on the existing idea of one point solution search meta-heuristics such as classical hill climbing and simulated annealing algorithms in which at each iteration; a new generated candidate solution is compared with a current one. However, the main strategy of LAHC is to memorize the fitness costs after each single move during the search process (C = LAHC list) with size (l). In LAHC, acceptance of the new solution depends on a comparison between the new candidate solution and previous solutions generated several iterations before. The "delay" in the comparing new a solution with its previous one motivates the name of this new one-point solution algorithm and also enables the usage of the simplest greedy acceptance mechanism. The algorithm accepts a candidate solution as long as their fitness cost is better than the one generated t iterations ago.

3.1   Adaptation of LAHC for the (OFA)

In this section, an adaptation of LAHC for tackling the OFA is presented. The adaptation process involves integrating the three different neighbourhood search within the operators of the LAHC in order to navigate the OFA search space effectively. It worthy of notice that the feasible region of the OFA search space is maintained during search and thus step in the process of adaptation of LAHC for OFA is provided in the next subsection.

Generate Initial Feasible OFA Solution. In this phase, the initial feasible solution to the OFA problem where the hard constraints are satisfied and resources are assigned to suitable rooms is generated based on the Peckish initialization procedure. The procedure is similar to what was utilized in [6,9] and the pseudocode of the initialization using the peckish procedure is provided in Algorithm 1.

Algorithm 1. Peckish initialization procedure

---
 while not all entities are assigned **do**
  $K = N/3$
  Select an unassigned entity j randomly
  Select a number of K rooms which satisfy $^1/_2 \times w_j \leq w_j \leq {}^3/_2 \times w_j$ randomly
  Select the best room from K rooms with the minimum penalty
  Assign the entity j to the best room
 end while

Neighbourhood Search. Neighbourhood search is the strategy employed to move the search towards neighbouring solutions from the existing one in solution search space. The three neighbourhood searches used in this study are:

1. *EntityRelocate − Neighbourhood*: This neighbourhood search randomly removes an entity x_i' from its current room and assigned to another available room randomly.
2. *EntitySwap − Neighbourhood*: This neighbourhood swaps the rooms of the two entities x_i' and x_j' which are randomly selected i.e. the room that is assigned to entity A is swapped with the room assigned to entity B randomly and vice versa.
3. *RoomInterchange − Neighbourhood*: This neighbourhood randomly selected two rooms, and all entities assigned to room A are interchanged with all entities assigned to room B.

Proposed LAHC for the OFA. The procedure of adapting the LAHC algorithm for the OFA problem is presented in this section. This is followed by the initialization of the parameters of the LAHC algorithm for the OFA. These parameters include MaxnoIteration that represents the number of iterations and l which is the table size (or penalty cost array). Similarly, the variables of the OFA that are also extracted from the dataset include entities, set of rooms and capacity of the room, the size of capacity needed by each entity and set of constraints (i.e. hard and soft). The OFA main decision variable is the entities where each entity could be assigned to a feasible resource in the OFA solution. A set of all feasible resources could be considered as the available range of such entities. Note that the feasible resources of each entity changes during the search of LAHC algorithm. The pseudocode of adapting the LAHC algorithm for the OFA is given in Algorithm 2.

The initial feasible solution to the OFA is generated randomly using the peckish initialization procedure and the fitness cost of the feasible solution is evaluated using Eq. 1. It is noteworthy that the feasibility region of the search space must be protected at the initial stage and during the search activities of the LAHC algorithm. The optimization of the feasible OFA solution is carried out with the aid of three randomly chosen neighbourhood searches in order to generate a new optimized solution. The cost value of the optimized solution is evaluated and if better than or equal to the penalty cost of the current solution and penalty cost is stored on the position of $v = i\ mod\ l$. Then the optimized solution replaces the current solution and thus the last element of the list l is removed and cost value of the new solution is stored at the beginning of the table list l. Furthermore, if the cost value of the optimized solution is better than the cost value of the best solution found so far, then the optimized solution is stored as the best solution. However, if the penalty cost of the optimized solution is not better than or equal to the penalty cost of the last element of the list l, then the optimized solution is rejected.

Algorithm 2. The pseudocode of the Proposed LAHC for OFA

1: Initialize the parameters of the LAHC and OFA
2: Initialize OFA feasible solution x using Algorithm 1
3: Evaluate the penalty cost $f(x)$
4: Output: Best OFA solution found x^*.
5: set $i = 0$
6: Set the *Iter = 1*
7: **for** $k \in (0 \ldots l - 1)$ $f(k)=f(x)$ **do**
8:     **while** $Iter \leq MaxIter)$ **do**
9:         /* Construct a candidate solution x' */
10:         $i = RND()$ {/}* RND generate a random integer number between 1-3 */
11:         **if** $(i == 1)$ **then**
12:             $x'_i = EntityRelocate(x_i)$
13:         **else**
14:             **if** $(i == 2)$ **then**
15:                 $x'_i = EntitySwap(x_i)$
16:             **else**
17:                 **if** $(i == 3)$ **then**
18:                     $x'_i = RoomInterchange(x_i)$
19:                 **end if**
20:             **end if**
21:         **end if**
22:         /* Penalty Cost Evaluation $f(x')$.
23:         $v = i \bmod l.$
24:         **if** $f(x'_i) \leq f(v_i)$ **then**
25:             $v_i = x'_i$
26:             **if** $f(v_i) \leq f(x^*)$ **then**
27:                 $x^* = v_i$
28:             **end if**
29:         **end if**
30:         $i= i + 1$
31:         $Iter = Iter + 1$
32:     **end while**
33: **end for**

---

4   Experimental Results and Discussions

The proposed LAHC algorithm is programmed with Microsoft Visual Basic.NET on Windows 8 platform on Intel® core i3-4005u CPU @1.70 GHz and 4 GB RAM and the results all instances of the dataset are obtained within computational time of 100 s. The benchmark instances of the datasets introduced by University of Nottingham and University of Wolverhampton are used in the evaluation of the performance of the proposed method. The characteristics of these datasets are provided in [9]. The parameter settings for LAHC is fixed as follows: Maxnon-Iter = 10000 while $l = 25$, which is adapted from [21].

4.1   Experimental Design

In order to study the sensitivity of using the three neighbourhood searches on the performance of the LAHC when adapted for tackling the OFA problem, this section provides an experimental design for the seven convergence cases (Cases 1–7) of different incorporations of these neighbourhood searches within the operator of the LAHC algorithm as shown in Table 1. Generally, all possible incorporations of these neighbourhood searches are studied separately.

Table 1. Experimental setup that effects of neighbourhood structure on LAHC algorithm for the OFP

Neighbourhood search	Case 1	Case 2	Case 3	Case 4	Case 5	Case 6	Case 7
EntityRelocate	✓	X	X	✓	✓	X	✓
EntitySwap	X	✓	X	✓	X	✓	✓
RoomInterchange	X	X	✓	X	✓	✓	✓

4.2   Experimental Results

The summary of the results obtained by the LAHC based on seven convergence cases are presented in Table 2. The values in Table 2 represent the penalty cost which is formulated in Eq. 1. Note that lowest value is the best. For each instance of the benchmark datasets, the best, average and worst of the 10 runs are recorded. The best result obtained by the proposed LAHC algorithm for each instance is highlighted in bold. As shown in Table 2, the proposed LAHC algorithm with integration of three neighbourhood structures (i.e., case 7) obtained best results in comparison with all other cases that integrated single or double neighborhood searches. Similarly, the results obtained by the case 5 shows that this version LAHC algorithm that incorporate *EntityRelocate* and *RoomInterchange* is able compete with case 7 which shows that incorporation of these neighbourhood searches within the component of the LAHC algorithm could aid the search navigation of the algorithm and thus improve the performance. Apparently, the performance of case 4 with *EntityRelocate* and *EntitySwap* neighbourhoods is closer to that case 5 in terms of the penalty cost in almost all the instances when compared with the remaining cases 1, 2, 3 and 6. Moreover, the results obtained by the LAHC algorithm is enhanced by lowering the penalty cost further with different incorporations of these neighbourhoods (i.e. Case 5 and 4). Finally, it can be seen that the results obtained prove that incorporation of two or more neighbourhoods with the operators of LAHC algorithm enhances the search capability of the proposed algorithm.

Table 2. Computational results of using different incorporation of neighbourhood searches on LAHC algorithm

Instances		Case 1	Case 2	Case 3	Case 4	Case 5	Case 6	Case 7
NOTT 1	Best	1219.45	1218.70	831.95	523.35	503.95	472.90	**393.80**
	Average	1344.91	1418.24	918.94	781.79	711.33	655.71	601.13
	Worst	1484.55	1652.15	1023.05	1302.25	1085.10	747.90	834.45
NOTT 1A	Best	1201.30	1207.05	845.75	503.90	486.70	629.35	**461.75**
	Average	1325.16	1345.55	963.47	770.63	614.93	764.13	665.39
	Worst	1391.35	1566.80	1136.85	1061.10	812.70	963.95	1067.85
NOTT 1B	Best	676.75	1069.30	744.60	375.65	363.20	499.35	**332.50**
	Average	731.43	1115.93	795.93	458.31	391.63	541.95	376.96
	Worst	826.05	1171.15	838.10	543.40	426.85	577.65	430.50
NOTT 1C	Best	994.30	647.65	496.75	379.40	427.70	418.45	**356.60**
	Average	1071.46	730.27	606.97	481.65	547.39	538.96	498.82
	Worst	1164.40	849.70	696.80	542.75	639.15	823.70	719.75
NOTT 1D	Best	495.05	425.15	335.00	295.65	305.65	328.50	**288.30**
	Average	538.49	487.44	362.05	338.46	346.23	411.42	332.59
	Worst	596.40	510.00	374.60	366.75	469.40	473.40	405.00
NOTT 1E	Best	579.90	740.70	448.90	226.70	144.90	252.90	**137.70**
	Average	633.27	851.33	507.85	316.35	232.57	291.92	203.15
	Worst	777.30	963.10	579.80	462.70	414.90	339.40	268.40
WOLVER 1	Best	697.25	634.19	634.37	634.19	634.19	634.19	**634.19**
	Average	756.00	660.77	653.23	637.16	690.20	686.53	678.51
	Worst	822.77	691.79	706.49	651.98	895.31	782.00	872.18
TRENT1	Best	4124.00	9801.00	9178.00	3200.00	3138.00	8574.00	**3072.00**
	Average	4642.71	9929.00	9327.33	3538.83	3658.50	8722.00	3608.83
	Worst	5497.00	10037.00	9396.00	4066.00	3929.00	8836.00	4245.00

4.3 Comparative Results

The computational results obtained by the LAHC algorithm using the two datasets are compared with results of other existing algorithms from the literature which include ABC-OFA [9], IPM-OFA [3], OFA-MP [3], HSA-OFA [6], SA-OFA [25], and HMHPB [25]. Note that the best results obtained by the different methods are presented in bold. The performance LAHC algorithm is better than other existing algorithms by achieving high quality solutions in all instances. The proposed LAHC algorithm achieved new results in two instances (i.e. NOTT 1 and NOTT 1E) of Nottingham dataset and had comparable performance in the remaining instances of the dataset. Similarly, the LAHC obtained best result WOLVER 1 as achieved by ABC-OFA, IMP-OFA and HSA-OFA methods (Table 3).

Table 3. The best results achieved by LAHC algorithm and other comparative techniques

INSTANCE	Proposed LAHC	ABC-OFA	IPM-OFA	OFA-MP	HAS-OFA	SA-OFA	HMHPB
NOTT1	**393.80**	425.50	=	=	539.35	543.70	482.20
NOTT1A	461.75	437.05	378.88	=	=	=	
NOTT1B	332.50	356.60	246.18	**243.28**	=	470.70	417.10
NOTT1C	356.60	324.20	**305.73**	**305.73**	=	342.50	315.40
NOTT1D	288.30	334.15	202.70	202.73	**200.10**	=	
NOTT1E	**137.70**	147.70	177.70	177.70	=	=	
WOLVER1	**634.19**	**634.19**	634.20	**634.19**	**634.19**	=	
TRENT1	3072.00	9885.00	=	=	=	2724.40	2531.40

5   Conclusion

This paper investigates a Late Acceptance Hill Climbing (LAHC), a recently proposed one-point meta-heuristic algorithm for solving Office-Space Allocation Problem. The LAHC algorithm is a variant of hill climbing optimizer that utilized an advanced acceptance criteria (i.e. late acceptance strategy) in order to prevent the algorithm from getting stuck in a local minimum. The OFA is a complex combinatorial optimization problem that involves allocating a set of resources limited to a set of spaces such that all resources are assigned to the require spaces subject to satisfying a set of constraints in order to achieve optimal utilization. The performance of the LAHC is evaluated using the datasets published by University of Nottingham and the University of Wolvehampton. The design of the experiment is intentionally made to test effects of the different combinations of these neighborhood searches on the performance of the proposed LAHC. The computational results proved that the LAHC incorporated with the three neighborhood searches is an effective technique for the OFA. It is noteworthy that a comparative evaluation with previous methods shows that proposed LAHC algorithm produced two new results, one best result and competitive results from the remaining instances. Finally, future research directions is to enhance the performance of the LAHC algorithm through the modifications and hybridizations with operators other metaheuristic algorithm.

References

1. Kellerer, H., Pferschy, U.: Cardinality constrained bin-packing problems. Ann. Oper. Res. **92**, 335–348 (1999)
2. McCollum, B.: A perspective on bridging the gap between theory and practice in university timetabling. In: Burke, E.K., Rudová, H. (eds.) PATAT 2006. LNCS, vol. 3867, pp. 3–23. Springer, Heidelberg (2007). https://doi.org/10.1007/978-3-540-77345-0_1
3. Ülker, Ö., Landa-Silva, D.: A 0/1 integer programming model for the office space allocation problem. Electron. Not. Discrete Math. **36**, 575–582 (2010)

4. Benjamin, C.O., Ehie, I.C., Omurtag, Y.: Planning facilities at the University of Missouri-Rolla. Interfaces **22**(4), 95–105 (1992)
5. Ritzman, L., Bradford, J., Jacobs, R.: A multiple objective approach to space planning for academic facilities. Manag. Sci. **25**(9), 895–906 (1979)
6. Awadallah, M.A., Khader, A.T., Al-Betar, M.A., Woon, P.C.: Office-space-allocation problem using harmony search algorithm. In: Huang, T., Zeng, Z., Li, C., Leung, C.S. (eds.) ICONIP 2012. LNCS, vol. 7664, pp. 365–374. Springer, Heidelberg (2012). https://doi.org/10.1007/978-3-642-34481-7_45
7. Burke, E.K., Cowling, P., Landa Silva, J.D., McCollum, B.: Three methods to automate the space allocation process in UK universities. In: Burke, E., Erben, W. (eds.) PATAT 2000. LNCS, vol. 2079, pp. 254–273. Springer, Heidelberg (2001). https://doi.org/10.1007/3-540-44629-X_16
8. Ülker, Ö., Landa-Silva, D.: Evolutionary local search for solving the office space allocation problem. In: 2012 IEEE Congress on Evolutionary Computation, CEC, pp. 1–8. IEEE (2012)
9. Bolaji, A.L., Michael, I., Shola, P.B.: Optimization of office-space allocation problem using artificial bee colony algorithm. In: Tan, Y., Takagi, H., Shi, Y. (eds.) ICSI 2017. LNCS, vol. 10385, pp. 337–346. Springer, Cham (2017). https://doi.org/10.1007/978-3-319-61824-1_37
10. Kirkpatrick, S., Gelatt, C.D., Vecchi, M.P., et al.: Optimization by simulated annealing. Science **220**(4598), 671–680 (1983)
11. Lopes, R., Girimonte, D.: The office-space-allocation problem in strongly hierarchized organizations. In: Cowling, P., Merz, P. (eds.) EvoCOP 2010. LNCS, vol. 6022, pp. 143–153. Springer, Heidelberg (2010). https://doi.org/10.1007/978-3-642-12139-5_13
12. Burke, E., Cowling, P., Silva, J.L.: Hybrid population-based metaheuristic approaches for the space allocation problem. In: Proceedings of the 2001 Congress on Evolutionary Computation, vol. 1, pp. 232–239. IEEE (2001)
13. Burke, E., Cowling, P., Landa Silva, J., Petrovic, S.: Combining hybrid metaheuristics and populations for the multiobjective optimisation of space allocation problems. In: Proceedings of the 2001 Genetic and Evolutionary Computation Conference, GECCO 2001, pp. 1252–1259 (2001)
14. Burke, E.K., Silva, J.D.L., Soubeiga, E.: Multi-objective hyper-heuristic approaches for space allocation and timetabling. In: Ibaraki, T., Nonobe, K., Yagiura, M. (eds.) Metaheuristics: Progress as Real Problem Solvers. ORCS, vol. 32, pp. 129–158. Springer, Boston (2005). https://doi.org/10.1007/0-387-25383-1_6
15. Burke, E.K., Bykov, Y.: A late acceptance strategy in hill-climbing for exam timetabling problems. In: PATAT 2008 Conference, Montreal, Canada (2008)
16. Yuan, B., Zhang, C., Shao, X.: A late acceptance hill-climbing algorithm for balancing two-sided assembly lines with multiple constraints. J. Intell. Manuf. **26**(1), 159–168 (2015)
17. Abuhamdah, A.: Experimental result of late acceptance randomized descent algorithm for solving course timetabling problems. Int. J. Comput. Sci. Netw. Secur. **10**(1), 192–200 (2010)
18. Verstichel, J., Berghe, G.V.: A late acceptance algorithm for the lock scheduling problem. In: Voß, S., Pahl, J., Schwarze, S. (eds.) Logistik Management, pp. 457–478. Springer, Heidelberg (2009)
19. Tierney, K.: Late acceptance hill climbing for the liner shippingfleet repositioning problem. In: Proceedings of the 14th EU/MEWorkshop, pp. 21–27 (2013)

20. Goerler, A., Schulte, F., Voß, S.: An application of late acceptance hill-climbing to the traveling purchaser problem. In: Pacino, D., Voß, S., Jensen, R.M. (eds.) ICCL 2013. LNCS, vol. 8197, pp. 173–183. Springer, Heidelberg (2013). https://doi.org/10.1007/978-3-642-41019-2_13

21. Bolaji, A.L., Bamigbola, A.F., Shola, P.B.: Late acceptance hill climbing algorithm for solving patient admission scheduling problem. Knowl.-Based Syst. **145**, 197–206 (2018)

22. Alzaqebah, M., Abdullah, S.: An adaptive artificial bee colony and late-acceptance hill-climbing algorithm for examination timetabling. J. Sched. **17**(3), 249–262 (2014)

23. Özcan, E., Bykov, Y., Birben, M., Burke, E.K.: Examination timetabling using late acceptance hyper-heuristics. In: IEEE Congress on Evolutionary Computation, CEC 2009, pp. 997–1004. IEEE (2009)

24. Jackson, W.G., Ozcan, E., Drake, J.H.: Late acceptance-based selection hyper-heuristics for cross-domain heuristic search. In: 2013 13th UK Workshop on Computational Intelligence, UKCI, pp. 228–235. IEEE (2013)

25. Landa-Silva, D., Burke, E.K.: Asynchronous cooperative local search for the office-space-allocation problem. INFORMS J. Comput. **19**(4), 575–587 (2007)

Applying an Iterated Greedy Algorithm to Production Programming on Manufacturing Environment Controlled by the PBC Ordering System

Fabio Molina da Silva$^{(\boxtimes)}$ and Roberto Tavares Neto

Federal University of Sao Carlos, Sao Paulo, Brazil
{fabio,tavares}@dep.ufscar.br

Abstract. Ordering systems are a mechanism used to program the flow of production orders into the manufacturing system. The correct usage and parametrization of such systems have a significant impact on the performance of the production. One of the well-succeed ordering systems available in the literature is the Period-batch control (PBC), that allows one to group the orders into different production periods, and program it into the planning horizon. This paper assumes a manufacturing system controlled by PBC. On this system, this paper considered two performance indicators: a primary goal is to minimize the total tardiness and the second goal is to minimize the idleness of the production system. Two approaches are implemented to solve this problem: a mixed-integer programming model and eight algorithms based on the Iterated Greedy method. Beyond finding good results when comparing to the ones found by the mathematical model approach, this paper also performs the Tardiness × Production Capacity on each algorithm.

Keywords: Periodic Batch Control · Iterated Greedy
Production planning and control · Scheduling

1 Introduction

The Periodic Batch Control (PBC) ordering system is a production programming and control that, given the demand specified by the Master Production Schedule (MPS), establish the required amount of components and materials. [1–7] emphasize that this ordering system is very common in production systems, especially the ones that seek modern manufacturing paradigms (such as lean manufacturing and quick response manufacturing). The efficiency of PBC is also strictly related to the generation of the MPS, especially on make-to-order systems [8].

This research was supported by CAPES, CNPq (407104/2016-0) and FAPESP (2010/10133-0).

© Springer Nature Switzerland AG 2019
M. J. Blesa Aguilera et al. (Eds.): HM 2019, LNCS 11299, pp. 191–199, 2019.
https://doi.org/10.1007/978-3-030-05983-5_15

The PBC considers that the production periods are equal, regardless of the amount of work allocated in each one. According [2], the set of policies defined by PBC guarantee that the processing of all orders occurs during the assigned production period.

Many manufacturing environments choose to adopt an implementation of the PBC ordering system instead of planning the shop floor schedule by using some algorithmic strategy to sequence and program each production order. This is mainly due to the idea of separate the set of orders into blocks of periods, where all the material transfer occurs only on the end of each block. Moreover, any process variability is easily absorbed by the capacity gaps introduced by the planner.

Even though many companies use PBC-based production planning, there is little research about how to allocate service orders into the production periods. Thus, this paper analyzes different strategies to perform this allocation. Are presented: a Mixed-Integer Programming model and an Iterated Greedy (IG) technique to elaborate the production plan of a make-to-order production system. The overall goal is to allow one to better use the existing production resources according to two objectives: firstly, to minimize total tardiness and, on a second stage, to minimize the idleness of the overall system. According to a literature review presented by [9], there is no previous use of the IG on the planning of the PBC ordering system.

2   The Iterated Greedy

The literature brings two major groups of algorithms to solve combinatorial problems: constructive algorithms that generate a new solution according to rules previously established and improvement algorithms, that, iteratively, seeks to generate better results than the current solution. As improvement algorithms, researches have been presenting strategies as Genetic Algorithms (GA - e.g., [3]), Particle Swarm Optimization (PSO - e.g. [4]), Ant Colony Optimization (ACO - e.g. [1]) and Iterated Greedy (IG - e.g. [8]). On the specific case of IG, a literature overview shows a set of applications, including algorithms to solve the Multiple Knapsack Problem (e.g. [2]), the Single Machine Scheduling (e.g. [9]) and the Flowshop Scheduling Problem (e.g. [8]).

The IG is composed of 3 main phases:

- An **Initialization Phase**, that generates an initial solution, usually based on a constructive heuristic. As an example, in the IG approach to a Permutational Scheduling Problem, [8] uses an insertion algorithm similar to the one used by the NEH algorithm [5].
- A **Destruction Phase**, that uses an implementation-specific strategy to remove elements from the current solution. Although previous researches have presented several destruction strategies (e.g., see [7]), a simple random sampling strategy is presented as a very suitable approach (e.g. [8]).
- A **Reconstruction Phase**, that uses some construction procedure to generate the new solution. Usually one apply a simple algorithm (e.g., [8] and [6] uses the insertion phase of the NEH algorithm).

Three further elements are also presented on the IG implementation found in the literature:

- An **acceptance criteria**, responsible for replacing the best solution found by the algorithm. The most straightforward strategy used to implement this feature is a pure greedy rule that the replacement occurs solely when the new solution found is better than the current best solution. To avoid it, researchers such as [6–8] suggest to use a simulated annealing-based rule.
- A **stopping criteria**, usually related to the number of destruction/reconstruction cycles, computational time spent or some stability measure of the final solution.
- A **local search** (optional) that can be used to assist the main algorithm to avoid local minimum solutions.

3   Problem Description

As mentioned before, the problem approached by this research considers a manufacturing system controlled by PBC, where a set of orders must be allocated into n production periods. The goal is to minimize the total tardiness of the orders and, considering this minimum total tardiness value, minimize the overall idleness of the production system. There is a constraint regarding the available production capacity at each production stage. For modeling issues, the orders that could not be produced on the available stages are programmed into an infinite-capacity dummy stage (the last one). Table 1 describe the symbols adopted into this research.

Table 1. Symbols used to describe the problem

Symbol	Definition
c	Number of production cycles (without the dummy stage)
P	Number of production orders
N	Number of production stages
HP	Size of the programming horizon ($HP = N + c$)
j	a production period ($j = 1, 2, ..., HP$)
i	an order ($i = 1, 2, ..., P$)
w	a productive stage ($w = 1, 2, ..., N$)
TP_{iw}	Processing time of order i on productive stage w
CP_{wj}	Productive capacity of stage w on period j
d_i	due date of order i

Table 2 presents the symbols used to identify a solution.

To accomplish the above statement goals, two MIP models were applied sequentially: the first one, composed by Eqs. 1–5, obtains the minimum value

Table 2. Solution variables

Symbol	Definition
A_i	Tardiness of order i
CO_{wj}	Idleness of stage w at period j
COR_{wj}	Real idleness of stage w at period j. This variable is included to assure that the objective function is not penalized when one have full idle periods
y_j	1 if any order is allocated into period j, 0 otherwise
x_{ij}	1 if order i is allocated to the period j, 0 otherwise

of $\sum_i A_i$; the second one, composed by Eqs. 6–9, minimizes the idleness of the system.

$$z_1 = \sum_i A_i \rightarrow min \tag{1}$$

$$\sum_j x_{ij} = 1 \qquad\qquad \forall i \tag{2}$$

$$\sum_{j \leq N-1} x_{ij} = 0 \qquad\qquad \forall i \tag{3}$$

$$\sum_i (TP_{iw} \cdot x_{ij}) + CO_{w(j-(N-w))} = CP_{w(j-(N-w))} \qquad \forall \begin{cases} w \\ j \geq N \end{cases} \tag{4}$$

$$A_i \geq j \cdot x_{ij} - d_i \qquad\qquad \forall \begin{cases} i \\ j \end{cases} \tag{5}$$

On this model, Eq. 1 if the objective function (minimize total tardiness); Eq. 2 assures that every order is allocated just to one period on PBC; Eq. 3 assures that no order will be allocated in a period before that it is required for its production; Eq. 4 establishes the capacity constrains of each period; Eq. 5 determine the values of the tardiness of each order.

$$z_2 = \sum_{w,j} COR_{wj} \rightarrow min \tag{6}$$

$$y_j \cdot M \geq \sum_i x_{ij} \qquad\qquad \forall j \tag{7}$$

$$COR_{w(j-N-w)} \geq CO_{w(j-N-w)} - M \cdot (1 - y_j) \qquad \forall \begin{cases} w \\ j \end{cases} \tag{8}$$

$$\sum_i A_i \leq z_1 \tag{9}$$

On this second model, the objective function is given by Eq. 6; Eqs. 7 and 8 assures that, if there is an order allocated on period j, $COR_{wj} = CO_{wj}$; otherwise, $COR_{wj} = 0$. Equation 9 assures that the minimum total tardiness found previously is maintained.

4   The IG Algorithm

The first stage on the design the IG algorithm is to define the initialization procedure. In this case, this paper developed the following algorithm: on the first stage the EDD (Earliest Due Date) is applied; if there are orders with the same due date, the second stage sequence those orders by the LPT (Longest Processing Time) rule. Once sequenced, the orders are allocated into the production period, assuring that the capacity constraints are respected.

The **destruction phase** is given by two different methods: (**D1**) a random-based removal of the existing orders from the current solution and **D2** a lookup function that removes the orders that use most of the production capacity.

The following methods were developed to the **reconstruction phase**: **R1** allocates a random order into the first available period (this is a variation of the bin-packing heuristic *Batch First Fit*); the method **R2** chooses the order with smaller sum of processing times and program on the first available period; the method **R3** selects the order with larger sum of processing times and program it on the first available period; **R4** sequences the set of removed jobs by the inverse of the total processing times and try to backward schedule the order to set the tardiness equal to zero. If it is not possible, a forward schedule is performed in such way that the minimum tardiness is achieved; finally, **R5** allocates the order with higher capacity utilization on the period with more idle capacity.

Using those procedures, 7 algorithms were developed, as shown in Table 3.

Table 3. Developed algorithms

Algorithm	Destruction	Reconstruction
H1A	D1	R5
H1B	D1	R1
H1C	D2	R5
H1D	D2	R2
H1E	D1	R3
H1F	D1	R4
H1G	D1	R2

5   Results and Analysis

Our literature review does not reveal any publicly available benchmarks to use in this work. Therefore, a set of test instances were generated using the following parameters: the number of orders were $n = \{30, 50, 100, 200$ or $300\}$; the number of production stages were given by $w = \{3, 5$ or $7\}$. The processing times were sampled from an uniform distribution $\{5, 20\}$. The due dates were sampled from an integer uniform distribution $\{w, w+j\}$. With no loss of generality, we assume the processing times and due dates as integers. The programming periods are 2 or 4 periods ahead. This procedure generates a set of 300 test instances.

The seven heuristics were coded in C++. The algorithms with stochastic components were executed 30 times and the better results were stored. The implementations were executed into a microcomputer i3 with 4GB RAM with Windows 7.

The instances were solved by the MIP model implemented in GAMS/CPLEX and by the heuristics. Two parameters were obtained: (i) a gap between the total tardiness found by CPLEX and each algorithm; (ii) a gap between the remaining capacity of every period found by CPLEX end by the algorithm. The average value of each gap is presented in Table 4 and Figs. 1 and 2.

Table 4. Average gap found (%)

	Total tardiness		Idleness	
	Average	Std. Dev.	Average	Std. Dev.
EDD	0.058	0.079	123.21	2022.02
H1A	0.058	0.079	123.18	2022.12
H1B	0.058	0.079	122.61	2017.71
H1C	0.055	0.076	114.69	1909.18
H1D	0.046	0.075	119.06	1972.08
H1E	0.055	0.077	94.53	1635.08
H1F	0.041	0.058	62.21	1042.89
H1G	0.058	0.078	120.4	2005.86

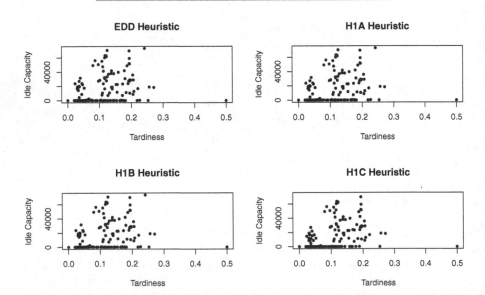

Fig. 1. Tardiness × Idle Capacity analysis for heuristics EDD, H1A, H1B and H1C

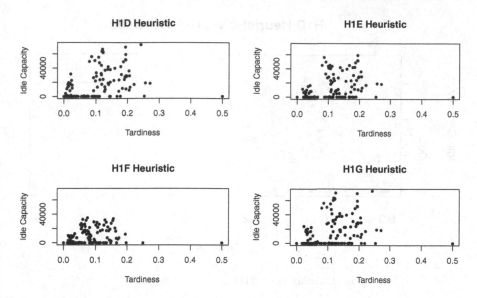

Fig. 2. Tardiness × Idle Capacity analysis for heuristics H1D, H1E, H1F and H1G

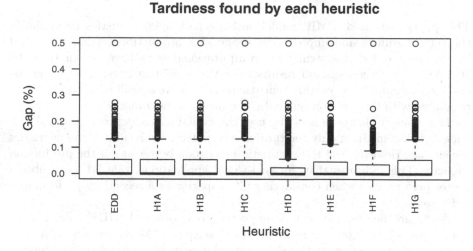

Fig. 3. Gap found by each heuristic

Figure 3 presents an analysis of the tardiness values found by each heuristic. As presented, the heuristics H1D and H1F could be able to find a better values on both total tardiness average and standard deviation.

Figure 4 compares the results of heuristics H1D and H1F. According this figure, heuristic H1F allows one to better use the production facility.

H1D Heuristic vs. H1F Heuristic

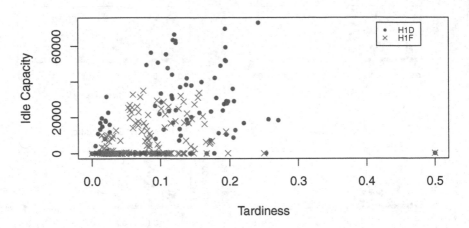

Fig. 4. Analysis of H1D and H1F heuristics

6 Conclusions

This paper presented a MIP model and a set of eight heuristics to minimize the total tardiness and improve the usage of a production system controlled by the PBC ordering system. Our computational tests have shown that the H1F heuristic presents better results than the remaining ones, considering the instances found. Moreover, this heuristic could achieve overall good results when compared with the solutions given by the mathematical model.

It was shown that each heuristic uses the available production capacity differently: when comparing only the tardiness objective, the results of the heuristics are similar. However, the H1F heuristic presents a better use of the production capacity. This indicates that, for real-world applications, rule H1F can obtain better performance when considering idle capacity as a secondary performance indicator.

As future developments, it is expected to improve the H1F rule, and to apply this solution procedure in practical scenarios. Moreover, one can realize the appeal to integrate the PBC with different decisions of the production system.

References

1. Dorigo, M., Stützle, T.: Ant Colony Optimization. MIT Press, Cambridge (2004)
2. Garcia-Martinez, C., Rodriguez, F., Lozano, M.: Tabu-enhanced iterated greedy algorithm: a case study in the quadratic multiple knapsack problem. Eur. J. Oper. Res. **232**(3), 454–463 (2014)
3. Goldberg, D.E.: Genetic Algorithms in Search, Optimization, and Machine Learning. Addison-Wesley Professional, Boston (1989)

4. Kennedy, J.: Particle swarm optimization. In: Sammut, C., Webb, G.I. (eds.) Encyclopedia of Machine Learning, pp. 760–766. Springer, Heidelberg (2011). https://doi.org/10.1007/978-0-387-30164-8

5. Nawaz, M., Enscore, E.E., Ham, I.: A heuristic algorithm for the m-machine, n-job flow-shop sequencing problem. Omega **11**, 91–95 (1983)

6. Pan, Q.-K., Wang, L., Zhao, B.-H.: An improved iterated greedy algorithm for the no-wait flow shop scheduling problem with makespan criterion. Int. J. Adv. Manuf. Technol. **38**(7–8), 778–786 (2008)

7. Rodriguez, F.J., Lozano, M., Blum, C., García-Martínez, C.: An iterated greedy algorithm for the large-scale unrelated parallel machines scheduling problem. Comput. Oper. Res. **40**(7), 1829–1841 (2013)

8. Ruiz, R., Stützle, T.: An iterated greedy heuristic for the sequence dependent setup times flowshop problem with makespan and weighted tardiness objectives. Eur. J. Oper. Res. **187**(3), 1143–1159 (2008)

9. Ying, K.-C., Lin, S.-W., Huang, C.Y.: Sequencing single-machine tardiness problems with sequence dependent setup times using an iterated greedy heuristic. Expert Syst. Appl. **36**(3), 7087–7092 (2009)

A Symbiotic Organisms Search Algorithm for Blood Assignment Problem

Prinolan Govender[1] and Absalom E. Ezugwu[2](✉)

[1] School of Mathematics, Statistics and Computer Science,
University of Kwazulu-Natal, Westville Campus, Private Bag X54001,
Durban 4000, South Africa
213535970@ukzn.ac.za
[2] School of Computer Science, University of KwaZulu-Natal,
King Edward Road, Pietermaritzburg Campus, Pietermaritzburg 3201,
KwaZulu-Natal, South Africa
ezugwua@ukzn.ac.za

Abstract. The demand for blood transfusion is considered a real world problem which is needed for various medical emergencies. The blood assignment problem was introduced to address this problem. The formulation of this problem stretches from managing critical blood shortage levels and blood unit expiration, to blood compatibility between donor and patients. Another contributing factor to the blood assignment problem, lies in the blood bank having to import additional blood units from external sources when supply cannot meet the demand. These challenges have serious consequences especially in the case where the demand for blood is very high. Taking these factors into consideration, this study implements a metaheuristic hybrid algorithm that combines symbiotic organisms search algorithm with the blood assignment policy in relation to the blood banks of South Africa. The aim of this study is to minimize blood product wastage with regards to expiration and importation, whilst maximizing product delivery to patients in need. In addition, this study also implements a unique way of generating randomized datasets based on social events relating to South Africa public holidays. The computational results indicate that the proposed hybrid algorithm performed well in minimizing blood importation, and experienced no form of expiration throughout the time period.

1 Introduction

Human blood inventory management is categorized by a string of influences which contributes to its efficiency and can complicate as time progresses [1]. In the past years, many aspects of blood management have been introduced and scrutinized in order to implement dynamic policies and strategies that would optimize the management process [2–4]. Blood is a perishable commodity with unique medical value to humans [2]. In accordance to the blood system, often referred to as the ABO system [3], there exist 8 blood types in humans. Blood compatibility plays a vital role in blood management, and distribution of such units [5]. Cases have risen where patients received incompatible blood types which resulted in blood clumping (also referred to as agglutination), which can be life threatening. The Blood Assignment Problem (BAP) can simply be

M. J. Blesa Aguilera et al. (Eds.): HM 2019, LNCS 11299, pp. 200–208, 2019.
https://doi.org/10.1007/978-3-030-05983-5_16

defined as an optimization process which efficiently assigns a supply of WB units to the daily demand of these units. The BAP has many underlying external components which contribute to the complexity of the problem. However, the main issue relates to the demand for Whole Blood (WB) units. Demand for WB units can be classified as either "expected" or "unexpected", and it is usually the unexpected component which causes issues with regards to WB unit supply.

In this paper, the possibility of improving the recently proposed Symbiotic Organisms Search (SOS) algorithm with an efficient blood assignment model to solve the BAP is investigated. The SOS Algorithm was first introduced in [12] to solve complex structural engineering design optimization problems. The SOS is capable of providing efficient and robust approach in exploiting and exploring large search space, more so, it has been employed to optimize a number of combinatorial optimization problems and have proved to be an efficient performer in that aspect [13–16]. The contribution of this paper involves the hybridization of SOS algorithm with a blood bank management policy. However, due to confidentiality issues, this study could not use real-world data and therefore, stochastic datasets, which were randomly generated were used to implement the proposed hybrid algorithm. This technique is further discussed in the later section of this paper. The policy also takes into account other contributing factors which could affect the management of blood products. These factors include: blood compatibility, the First-In-First-Out (FIFO) issuing system, expiration of WB units and importation of additional WB units from external sources.

2 Methodology

Every day the demand for WB units must be met. If the daily supply is greater than or equal to the daily demand, then the supply is distributed accordingly and the demand is considered as satisfied. However, if the daily demand exceeds supply, this then initiates other processes that would meet the desired level for demand. First, the blood bank must check for compatible blood types and use only the remaining units from the blood groups (each blood type is expected to fulfil their respective demand first). If pulling from additional blood units still has not satisfied the demand, then the blood bank must import additional units from external sources in order to satisfy the request. Overall the BAP can be summarized into 4 major components: Supply, Demand, Importation and Expiration. The proposed BAP objective function aims to minimize the combination of both importation and expiration over a finite period of time for all the blood types.

Generating Demand and Supply: Due to confidentiality issues, it was not possible to use real-world datasets in this study. Instead, a randomly generating datasets which utilises South African social trends based on monthly statistics was used. By incorporating monthly holidays as well as terms from educational institutions [9], it was theoretically possible to create unique percentage bounds and allocate them to each month, which in turn reduces unpredictability when generating dataset values. Reports have previously indicated that levels of drunken driving increases during the Easter period [8], thus blood banks tend to stock-pile blood products for precautionary measures during these public holidays. In terms of generating percentage bounds for supply, there were no significant events that occurred during a standard South African year, therefore the percentage bounds will be set between 25–75%. In South Africa, months like December (with many public holidays and closed schooling institutes) would experience much higher levels of WB units demand, unlike February, which has no form of public holidays.

Therefore, if we denote following as:

A: Represent the initial volume in a blood bank
d: Represent a day
m: Represent a month
b: Represent a blood type
Bu: Represent the upper percentage bound
Bl: Represent the lower percentage bound
rng: Represents a random generator

$$Supply_b \text{ or } Demand_b = A \cdot \left(rng(B_U - B_l)_m\right) \tag{1}$$

From Eq. 1, the supply or demand was generated by randomly selecting a percent between the upper and lower bounds depending on the month the system was currently in This was then multiplied by the initial volume in the blood bank to generates a value for supply or demand. The process for generating supply has an additional step, which involves adding the previous days' remainder (as long as the remainder was greater than 0). However, if the system was in the first day, then remainder was taken as 0.

3   Symbiotic Organisms Search Algorithm

The SOS is an algorithm which emulates the interactive behaviour of creatures within nature [10], with a notable advantage of having no specific parameter tuning, which decreases time in order to achieve good results [11]. The SOS can be divided into 3 main optimization phases namely: Mutualism, Commensalism and Parasitism. Each of these phases tries to modify the chosen individual(s) with the hopes of obtaining a more improved solution.

Mutualism: Organisms interact with each other in a way that benefits both parties. Let X_i and X_j represent 2 random individuals within a population, MV represent the Mutual Vector, X_{best} represent the organism with the best advantage, and BF represent the benefit factor. The mutualism phase is presented using the following equations.

$$X_{inew} = X_i + rand(0, 1) \cdot (X_{best} - MV \cdot BF1) \tag{2}$$

$$X_{jnew} = X_j + rand(0, 1) \cdot (X_{best} - MV \cdot BF2) \tag{3}$$

Where:

$$MV = (X_i + X_j)/2 \qquad (4)$$

The value obtained from $(X_{best} - MV)$ tries to increase survival in the population, with all improved individuals replacing the original individuals. The values of the benefit factors BF1 and BF2 are determined randomly using Eqs. 5 and 6.

$$BF_1 = (1 + round(rand(0, 1)), \quad |rand \in [0, 1] \qquad (5)$$

$$BF_2 = (1 + round(rand(0, 1)), \quad |rand \in [0, 1] \qquad (6)$$

Where round and rand are MATLAB function. The round function rounds up generated values to the nearest whole number and the rand function generates random number.

Commensalism: In this phase, the individual organism interacts with each other in a way that results in one organism benefiting without harming the other organism. Selection of two organisms is done randomly from the population, and have their fitness values evaluated. The fitter individual is labelled as X_i and the inferior individual is labelled as X_j.

$$X_{inew} = X_i + rand(-1, 1).(X_{best} - X_j) \qquad (7)$$

$$X_i \text{ benefits from } X_j \text{ by means of } (X_{best} - X_j) \qquad (8)$$

Parasitism: In this phase, the organisms interact with each other in a way that benefits one organism (parasite) whilst harming the other organism (host). To evaluate a form of parasitism for the BAP, two individuals from a population are randomly selected, with each of its fitness values evaluated similar to the commensalism phase. Following the evaluation, the fitter individual is labelled as the parasite, and the inferior as the host. The parasite then swaps segments of its representation with the host only if the value (from the host) improves its original solution.

Solution Representation. As mentioned previously, there are eight (8) different blood types for humans, which is donate here as the SOS organisms. Using this information, it is possible to extrapolate a solution representation pattern for an individual organism within a population of ecosystem. The individual organisms are finite with 8 segments, and each segment is represented with a specific blood type capable of containing a value of type double. Figure 1 depicts the individual organism used in the SOS

Organism 1	Organism 2	Organism 3	Organism 4	Organism 5	Organism 6	Organism 7	Organism 8
A⁺	A⁻	B⁺	B⁻	AB⁺	AB⁻	O⁺	O⁻

Fig. 1. Solution representation for the BAP

algorithm with each segment representing an individual blood type and a specific numerical value.

The Algorithm 1 below represents the general implementation of the improved SOS algorithm. Furthermore, Algorithm 1 also exhibits the functioning of the parasitism phase were the parasite analyses the host and swaps segments if the host contains a better value than itself.

Algorithm 1: Symbiotic organism search algorithm

1: $n = Ecosystem\ size$
2: $Par = Parasite$
3: **Begin**
4: Generate initial population of blood types X={X$_1$, X$_2$, ..., X$_n$}, and evaluate its fitness
5:   **While** stopping criteria is false
6:     **For** $i = 1$: n
7:       Calculate fitness of each individual organism (blood types)
8:       X_{best} = individual with lowest fitness
9:     **End For**
10:    //Implement the three SOS interaction phases
11:    Mutualism phase mentioned
12:    Commensalism phase
13:    Parasitism phase
14:    {
15:      for i = 1: $parasite$.length
16:        if ($parasite$ [i] is not equal to $demand$ [i])
17:          store $diff1 = demand$ [i] − $host$ [i]
18:          store $diff2 = demand$ [i] − $parasite$ [i]
19:        end if
20:        if ($diff1 < diff2$)
21:          swap host and parasite segments.
22:        end if
23:        if ($diff1 \geq diff2$)
24:            do not replace value.
25:        end if
26:    }
27:      **If** (fitness (Par) < fitness (X_{best}))
28:        $X_{best} = Par$
16.   **End While**
17: **Return** X_{best}
18: **End**

It is noteworthy to mention here that because the basic SOS algorithm was originally designed for solving continuous optimization problem and the BAP being a combinatorial optimization problem, a random permutation process of using a modulus function given by $u = \lfloor x + k \rfloor$ mod m was used to convert the solution x to an integer u, where, k and $m > 0$ are integers.

4   Experimental Setting and Dataset Generation

The SOS algorithm was implemented on Intel core i5 CPU with 2.5 GHz and 4 GB RAM and Windows 10.0 Operating system, while the implementation software is Java. For each dataset, the algorithm was run for 1000 iterations, using population size of 50 organisms. Previous studies resorted to generating datasets by means of incorporating fixed percentage bounds between the ranges of 25 to 75%. This was used to generate both demand and supply. This study tries to minimize the unpredictability associated

Table 1. Study datasets and percentage bounds pertaining to each dataset.

Dataset	Initial blood volume	Demand bounds (%)	Supply bounds (%)	Description
1	500	25–75	25–75	Experimental control- Used in [2, 6, 7]
2	500	SAGV	25–75	South African statistics for generating the value for demand
3	500	75–100	25–50	Examines: Demand > Supply
4	500	25–50	75–100	Examines: Supply > Demand
5	5000	25–75	25–75	Dataset 1 with larger volume
6	5000	SAGV	25–75	Dataset 2 with larger volume

with stochastic dataset when generating demand values, by incorporating public holidays and schooling terms based on the South Africa vacation trends, referred here as South African generated values (SAGV). Table 1 represents the datasets used in this study, as well as the percentage bounds pertaining to each dataset.

4.1   Results and Discussion

In Table 2, the average computational results achieved by the SOS implementation for the BAP in accordance to each dataset described in Table 1. The significant components used to analyse the performance of the hybrid SOS algorithm are the two variables which create the objective function, namely the expiration and importation of WB units. A solution was found if the supply for the day matches the daily demand. However, due to the way in which supply for a day was generated such that the previous days' remainder was added to the newer influx of donations, stock-piling can occur. Figure 2 also gives an indication as to when the stock-piling event occurs. It is important to mention that the quicker this happens, the fewer would be the importation of the WB unit. Figure 2, represent the line graphs plots over a 365-day period for each dataset used to test the hybrid implementation of the SOS algorithm that was combined with the blood bank assignment policy.

In Table 2 and Fig. 2, the results indicate that the hybrid SOS algorithm coupled with the blood management policy achieved good results in terms of having very low importation levels and no form of expiration across any of the datasets. The results also show that stock-piling occurs at early period within the time frame which supports low

Table 2. Averages obtained for each dataset per blood type using the hybrid SOS algorithm to work the BAP

Dataset		A^+	A^-	B^+	B^-	AB^+	AB^-	O^+	O^-
1	Supply	192.81	88.67	78.67	35.41	15.27	6.36	131.83	87.8
	Demand	40	6.25	15	2.5	3.75	1.25	48.75	8.75
	Import	0	0	0.08	0.01	0.28	0.01	0.02	0
	Expiry	0	0	0	0	0	0	0	0
2	Supply	222.87	43.48	76.14	24.06	19.19	9.34	294.67	52.15
	Demand	19.08	2.98	7.15	1.19	1.79	0.6	23.25	4.17
	Import	0.11	0	0.03	0	0.01	0	0.08	0
	Expiry	0	0	0	0	0	0	0	0
3	Supply	68.9	28.51	27.04	11.63	3.08	2.4	72.98	30.71
	Demand	37.16	5.81	13.93	2.32	3.48	1.16	45.29	8.13
	Import	1.2	0.01	0.39	0.01	1.51	0.1	0.68	0
	Expiry	0	0	0	0	0	0	0	0
4	Supply	3419.58	565.67	1308.53	209.98	326.2	98.9	4172.27	746.18
	Demand	18.84	2.94	7.07	1.18	1.77	0.59	22.96	4.12
	Import	0	0	0	0	0	0	0	0
	Expiry	0	0	0	0	0	0	0	0
5	Supply	2404.66	1184.73	882.15	438.4	238.29	86.86	1639.28	1220.06
	Demand	397.86	62.17	149.2	24.87	37.3	12.43	484.89	87.03
	Import	0	0	0	0	0.62	0	0	0
	Expiry	0	0	0	0	0	0	0	0
6	Supply	5157.35	914.22	2439.9	430.83	549.99	206.36	7621.29	1389.61
	Demand	354.62	55.41	132.98	22.16	33.25	11.08	432.2	77.57
	Import	0	0	0	0	0.18	0	0	0
	Expiry	0	0	0	0	0	0	0	0

importation levels, but opens the system to possible expiration, however this phenomenon did not occur due to the 30 shelf life of WB units as well as the First-In-First-Out (FIFO) issuing system.

The studies conducted in [3, 4, 6, 7] used constant percentage bounds ranging between 27–75% in order to generate values for demand. These bounds were used across the entire testing range of 365 days. The current study allocates specific percentage ranges to each month with the aim of generating more accurate demand levels in accordance to the South African monthly schooling terms and public holidays. Ideally, the best source of generating demand percentage bounds would preferably be statistics based on actual demand for WB units within South Africa, however, these statistics was not available.

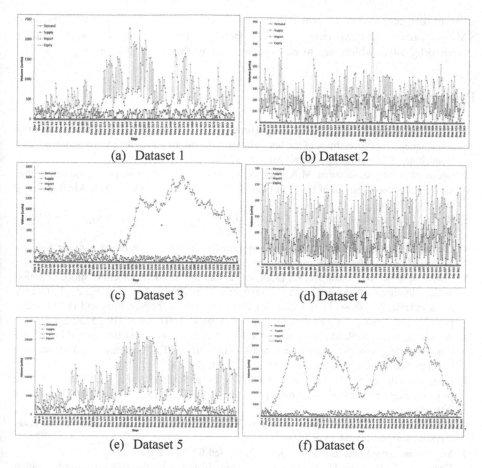

(a) Dataset 1   (b) Dataset 2

(c) Dataset 3   (d) Dataset 4

(e) Dataset 5   (f) Dataset 6

Fig. 2. SOS implementation of all datasets over a period of 365 days

5 Conclusion

This paper presents a hybrid metaheuristic algorithm, which combines symbiotic organism search algorithm with a blood assignment heuristic method to solve the BAP. Based on the results discussed in this paper, it is obvious that the hybrid SOS implementation was able to provide good quality solutions to the BAP. The implementation had very low amounts of importation, and no form of expiration when subjected to any of the datasets. The low importation levels can be attributed to the effects of using the bottom-up technique which promoted the use of compatible blood types, and the lack of expiration can be linked to the FIFO issuing system. Using dataset 1 as a control, it was possible to establish how effective the demand generation was in relation to using SAGV. Dataset 2 experienced a 42.5% overall decrease in the total levels for importation, whilst dataset 5 and 6 used the same percentage bounds as dataset 1 and 2, but had a much larger initial volume of WB units. Dataset 6 (which used SAGV for generating demand), also experienced a lower importation level by

44%. In relation to dataset 1 and 2 in terms of demand generation, dataset 2 used the SAGV approach, and experienced a 52% decrease in overall demand, whilst supply increased by 14%, which was to be expected due to the act of stock-piling.

References

1. Hesse, S., Coullard, C., Daskin, M., Hurter, A.: A case study in platelet inventory management. In: Proceedings of the 6th Industrial Engineering Research Conference, p. 801-6. Institute of Industrial Engineers, Atlanta (1997)
2. Olusanya, M.O., Arasomwan, M.A., Adewumi, A.O.: Particle swarm optimization algorithm for optimizing assignment of blood in blood banking system. Comput. Math. Methods Med. **2015**, 12 (2015)
3. Reid, M.E., Lomas-Francis, C., Olsson, M.L.: The Blood Group Antigen Factsbook. Academic Press, London (2012)
4. Charpin, J.P., Adewumi, A.O.: Optimal assignment of blood in a blood banking System. Technical report, Mathematics in Industry Study Group (MISG) (2011)
5. Baş, S., Carello, G., Lanzarone, E., Ocak, Z., Yalçındağ, S.: Management of blood donation system: literature review and research perspectives. In: Matta, A., Sahin, E., Li, J., Guinet, A., Vandaele, N.J. (eds.) Health Care Systems Engineering for Scientists and Practitioners, pp. 121–132. Springer, Cham (2016). https://doi.org/10.1007/978-3-319-35132-2_12
6. Adewumi, A.O., Budlender, N., Olusanya, M.O.: Optimizing the assignment of blood in a blood banking system: some initial results. In: 2012 IEEE Congress on Evolutionary Computation (CEC), 10 June 2012, pp. 1–6. IEEE (2012)
7. Igwe, K., Olusanya, M., Adewumi, A.O.: On the performance of GRASP and dynamic programming for the blood assignment problem. In: GHTC 2013, 20 October 2013, pp. 221–225 (2013)
8. https://www.news24.com/SouthAfrica/News/easter-road-death-toll-up-by-14-from-last-year-20180417. Accessed 04 June 2018
9. http://www.schoolterms.co.za/2018.html. Accessed 04 June 2018
10. Cheng, M.Y., Prayogo, D.: Symbiotic organisms search: a new metaheuristic optimization algorithm. Comput. Struct. **15**(139), 98–112 (2014)
11. Cheng, M.Y., Prayogo, D., Tran, D.H.: Optimizing multiple-resources leveling in multiple projects using discrete symbiotic organisms search. J. Comput. Civil Eng. **30**(3), 04015036 (2015)
12. Cheng, M.Y., Prayogo, D.: Symbiotic organism search: a new metaheuristic optimization. Comput. Struct. **139**, 98–112 (2014)
13. Ezugwu, A.E., Adewumi, A.O., Frîncu, M.E.: Simulated annealing based symbiotic organisms search optimization algorithm for traveling salesman problem. Expert Syst. Appl. **77**, 189–210 (2017)
14. Tran, D.H., Cheng, M.Y., Prayogo, D.: A novel multiple objective symbiotic organisms search (MOSOS) for time–cost–labor utilization tradeoff problem. Knowl.-Based Syst. **94**, 132–145 (2016)
15. Ezugwu, A.E., Aderemi, A.O.: Discrete symbiotic organisms search algorithm for travelling salesman problem. Expert Syst. Appl. **87**, 70–78 (2017)
16. Ezugwu, A.E., Adeleke, O.J., Viriri, S.: Symbiotic organisms search algorithm for the unrelated parallel machines scheduling with sequence-dependent setup times. PLoS ONE **13**(7), e0200030 (2018)

Author Index

Printed in the United States
By Bookmasters